Computational Methods and Production Engineering

Related titles

Industrial Tomography
(ISBN: 9781782421184)

Computational Materials Engineering
(ISBN: 9780123694683)

26th European Symposium on Computer Aided Process Engineering
(ISBN: 9780444634283)

24th European Symposium on Computer Aided Process Engineering
(ISBN: 9780444634344)

Woodhead Publishing Reviews: Mechanical Engineering Series

Computational Methods and Production Engineering

Research and Development

Edited by

J. Paulo Davim

University of Aveiro, Aveiro, Portugal

Woodhead Publishing is an imprint of Elsevier
The Officers' Mess Business Centre, Royston Road, Duxford, CB22 4QH, United Kingdom
50 Hampshire Street, 5th Floor, Cambridge, MA 02139, United States
The Boulevard, Langford Lane, Kidlington, OX5 1GB, United Kingdom

Library of Congress Cataloging-in-Publication Data
A catalog record for this book is available from the Library of Congress

British Library Cataloguing-in-Publication Data
A catalogue record for this book is available from the British Library

ISBN: 978-0-85709-481-0 (print)
ISBN: 978-0-85709-482-7 (online)

For information on all Woodhead publications visit our website at https://www.elsevier.com/books-and-journals

Publisher: Glyn Jones
Acquisition Editor: Glyn Jones
Editorial Project Manager: Lindsay Lawrence
Production Project Manager: Stalin Viswanathan
Cover Designer: Matthew Limbert

Typeset by SPi Global, India

Contents

List of contributors **ix**
About the editor **xi**
Preface **xiii**

1 Parallel direct solver for finite element modeling of manufacturing processes **1**
C.V. Nielsen, P.A.F. Martins
1.1 Introduction **1**
1.2 Brief review of standard gauss elimination **3**
1.3 Structure of the parallel direct solver **6**
1.4 Test cases and evaluation parameters **14**
1.5 Results and discussion **16**
1.6 Conclusions **25**
Appendix **26**
Acknowledgments **30**
References **30**

2 Optimal inspection/actuator placement for robust dimensional compensation in multistage manufacturing processes **31**
J.V. Abellán-Nebot, I. Peñarrocha, E. Sales-Setién, J. Liu
2.1 Introduction **31**
2.2 State-space approach for modeling multistage assembly processes **33**
2.3 Feed-forward predictive control **36**
2.4 Optimal inspection/actuator placement for feed-forward control **39**
2.5 Case study **43**
2.6 Conclusions **46**
Appendix **47**
References **50**

3 Numerical optimization strategies for springback compensation in sheet metal forming **51**
A. Maia, E. Ferreira, M.C. Oliveira, L.F. Menezes, A. Andrade-Campos
3.1 Introduction **51**
3.2 Springback compensation strategies **52**
3.3 Modeling strategies **55**
3.4 Evaluation strategies **60**
3.5 Optimization algorithms **62**

3.6 Parameterization strategies **62**
3.7 Case study: U-rail **63**
3.8 Results and discussion **69**
3.9 Conclusions **79**
Acknowledgments **79**
References **79**

4 Finite element modeling of hot rolling: Steady- and unsteady-state analyses **83**
Matruprasad Rout, Surjya K. Pal, Shiv B. Singh
4.1 Introduction **83**
4.2 Thermomechanical analysis of hot rolling: An overview **86**
4.3 Work-roll and workpiece interface behavior **91**
4.4 Constitutive equation for material model **94**
4.5 Basic steps of FEM **95**
4.6 Different approaches in FEM **97**
4.7 Solution methods **98**
4.8 Steady- and unsteady-state analyses of hot rolling **99**
4.9 Concluding remarks **119**
References **119**

5 Numerical modeling methodologies for friction stir welding process **125**
Rahul Jain, Surjya K. Pal, Shiv B. Singh
5.1 Introduction to FSW **125**
5.2 Modeling of FSW: Requirement and complexities **127**
5.3 General steps for modeling a process **128**
5.4 Modeling of FSW with Lagrangian analysis **141**
5.5 Modeling of FSW with Eulerian analysis **153**
5.6 Modeling of FSW with coupled Eulerian-Lagrangian (CEL) analysis **158**
5.7 Comparison of modeling methods **164**
5.8 Conclusion **164**
References **165**

6 Modeling of hard machining **171**
N.E. Karkalos, A.P. Markopoulos
6.1 Introduction to hard machining **171**
6.2 Numerical modeling of hard machining **177**
6.3 Soft computing and statistical methods modeling of hard machining **186**
6.4 Conclusions **193**
References **193**

7 Multiresponse optimization in wire electric discharge machining (WEDM) of HCHCr steel by integrating response surface methodology (RSM) with differential evolution (DE) **199**
V.N. Gaitonde, M. Manjaiah, S. Maradi, S.R. Karnik, P.M. Petkar, J. Paulo Davim
7.1 Introduction **199**
7.2 Experimental work **201**
7.3 Response surface methodology **204**
7.4 Results and discussion **205**
7.5 Differential evolution (DE) optimization **216**
7.6 Conclusions **219**
References **220**

Index **223**

List of contributors

J.V. Abellán-Nebot Universitat Jaume I, Castelló de la Plana, Spain

A. Andrade-Campos University of Aveiro, Aveiro, Portugal

E. Ferreira University of Aveiro, Aveiro; University of Coimbra, Coimbra, Portugal

V.N. Gaitonde B.V.B. College of Engineering and Technology, Hubli, India

Rahul Jain Indian Institute of Technology, Kharagpur, India

N.E. Karkalos National Technical University of Athens, Athens, Greece

S.R. Karnik B.V.B. College of Engineering and Technology, Hubli, India

J. Liu The University of Arizona, Tucson, AZ, United States

A. Maia University of Aveiro, Aveiro, Portugal

M. Manjaiah GeM Lab, Nantes, France

S. Maradi B.V.B. College of Engineering and Technology, Hubli, India

A.P. Markopoulos National Technical University of Athens, Athens, Greece

P.A.F. Martins Universidade de Lisboa, Lisbon, Portugal

L.F. Menezes University of Coimbra, Coimbra, Portugal

C.V. Nielsen Technical University of Denmark, Lyngby, Denmark

M.C. Oliveira University of Coimbra, Coimbra, Portugal

Surjya K. Pal Indian Institute of Technology, Kharagpur, India

J. Paulo Davim University of Aveiro, Aveiro, Portugal

I. Peñarrocha Universitat Jaume I, Castelló de la Plana, Spain

P.M. Petkar B.V.B. College of Engineering and Technology, Hubli, India

Matruprasad Rout Indian Institute of Technology, Kharagpur, India

E. Sales-Setién Universitat Jaume I, Castelló de la Plana, Spain

Shiv B. Singh Indian Institute of Technology, Kharagpur, India

About the editor

J. Paulo Davim received his PhD in Mechanical Engineering in 1997, MSc degree in Mechanical Engineering (materials and manufacturing processes) in 1991, Dipl.-Ing Engineer's degree (5 years) in Mechanical Engineering in 1986, from the University of Porto (FEUP), the Aggregate title (Full Habilitation) from the University of Coimbra in 2005 and a DSc from London Metropolitan University in 2013. He is Eur Ing by FEANI-Brussels and Senior Chartered Engineer by the Portuguese Institution of Engineers with a MBA and Specialist title in Engineering and Industrial Management. Currently, he is Professor at the Department of Mechanical Engineering of the University of Aveiro, Portugal. He has more than 30 years of teaching and research experience in Manufacturing, Materials, and Mechanical Engineering with special emphasis in Machining and Tribology. He has also interest in Management/Industrial Engineering and Higher Education for Sustainability/ Engineering Education. He has received several scientific awards. He has worked as evaluator of projects for international research agencies as well as examiner of PhD thesis for many universities. He is the Editor in Chief of several international journals, Guest Editor of journals, books Editor, book Series Editor, and Scientific Advisory for many international journals and conferences. Presently, he is an Editorial Board member of 30 international journals and acts as reviewer for more than 80 prestigious Web of Science journals. In addition, he has also published as editor (and coeditor) of more than 100 books and as author (and coauthor) of more than 10 books, 80 book chapters, and 400 articles in journals and conferences (more than 200 articles in journals indexed in Web of Science/h-index 37+ and SCOPUS/h-index 45+).

J. Paulo Davim
University of Aveiro, Aveiro, Portugal

Preface

Production engineering is a branch of engineering that involves "the design, development, implementation, operation, maintenance, and control of all processes in the manufacture of a product." It is an interdisciplinary subject requiring the collaboration of individuals trained in manufacturing engineering, industrial engineering, product design, management, etc.

Recently, there has been increased interest in developing computational methods to be applied in production engineering. Consequently, in production engineering, computational methods have achieved several applications, namely, modeling manufacturing processes, monitoring and control, parameters optimization and computer-aided process planning, etc.

This research book aims to provide information on computational methods and production engineering for modern industry. The initial chapter of the book provides parallel direct solver for finite element modeling of manufacturing processes. Chapter 2 is dedicated to optimal inspection/actuator placement for robust dimensional compensation in multistage manufacturing processes. Chapter 3 presents numerical optimization strategies for springback compensation in sheet metal forming. Chapter 4 covers finite element modeling of hot rolling (steady- and unsteady-state analyses). Chapter 5 is dedicated to numerical modeling methodologies for friction stir welding process. Chapter 6 contains information on modeling of hard machining. Finally, the last chapter of the book is dedicated to multiresponse optimization in wire electric discharge machining (WEDM) of HCHCR steel by integrating response surface methodology (RSM) with differential evolution (DE).

The present book can be used as a research book for final undergraduate engineering course or as a topic on computational methods and production engineering at the postgraduate level. This book can serve as reference for academics, researchers, manufacturing, mechanical and industrial engineers, as well as professionals in production engineering. Also, this book presents scientific interest for industry, centers of the research, laboratories, and universities throughout the world.

The editor acknowledges Woodhead/Elsevier for this opportunity and for their professional support. Finally, I would like to thank all the chapter authors for their availability for this work.

J. Paulo Davim
University of Aveiro, Aveiro, Portugal

Parallel direct solver for finite element modeling of manufacturing processes

1

C.V. Nielsen, P.A.F. Martins†*
*Technical University of Denmark, Lyngby, Denmark, †Universidade de Lisboa, Lisbon, Portugal

1.1 Introduction

The central processing unit (CPU) time is of paramount importance in finite element modeling of manufacturing processes. Because the most significant part of the CPU time is consumed in solving the main system of equations resulting from finite element assemblies, different approaches have been developed to optimize solutions and reduce the overall computational costs of large finite element models.

The simplest approach is to apply faster solution techniques by replacing direct equation solvers by iterative equation solvers. There are various types of iterative solvers but the conjugate gradient (CG) iterative solvers proposed by Lanczos (1952) and Hestenes and Stiefel (1952) are among the simplest and more widely used in finite element computer programs.

Despite the advantages of CG iterative solvers, there are two major concerns (and challenges) related to its utilization in finite element modeling of manufacturing processes. First, precision is lost compared to that of direct solvers, because final accuracy depends on the threshold value utilized for accepting the solution, which inevitably results from a compromise between the desired accuracy and the required CPU time. In case small inaccuracies accumulate during finite element modeling of a manufacturing process involving a large number of solution steps, they may lead to poor satisfaction of the boundary conditions and to inaccuracies in fulfilling symmetry conditions. For instance, a zero displacement associated with a symmetry line (or plane) may be computed as a very small nonzero displacement creating problems in the overall final accuracy of the numerical simulation. In problems involving contact the earlier mentioned problems may, for larger threshold values, also disturb the contact algorithms, eventually leading to penetration in contact pairs.

The second drawback results from the fact that CG iterative solvers suffer from low robustness and applicability in case of ill-conditioned equation systems. In fact, iterative solvers have been reported unstable when dealing with ill-conditioned equation systems, whereas direct solvers have proved to be more robust (Farhat and Wilson, 1988). This is particularly relevant for finite element formulations that make use of penalties (i.e., very large numbers) to impose material incompressibility

Computational Methods and Production Engineering. http://dx.doi.org/10.1016/B978-0-85709-481-0.00001-X

and/or to handle contact between different objects (e.g., deformable vs. deformable objects and/or deformable vs. rigid objects) and for simulations with large rigid body motion (e.g., when a preform is settling into a metal forming die undergoing very little deformation).

Among various improvements to optimize convergence and to improve the CPU time of CG iterative solvers, preconditioning and parallelization have been the most widely used techniques (Meijerink and van der Vorst, 1977).

A more complex approach to optimize solutions and reduce the overall computational cost of large finite element models has been achieved by decomposition of a finite element model into subdomains that are simultaneously solved by means of parallel computation (Kim and Im, 2003). The interface nodes between subdomains require communication between different processors and, therefore its overall number should be minimized in order to avoid computational bottlenecks. Farhat (1988) and Al-Nasra and Nguyen (1992) were among the first researchers to explore the trade-off between the number of interface nodes and subdomain size and to propose algorithms for optimal decompositions.

Parallelization of direct solvers is usually not considered as an alternative to parallelization of CG iterative solvers because it requires large amount of data communication between processors and also because it is considered more tedious and difficult to implement in existing finite element computer programs. However, parallelization of direct solvers may be an appropriate solution for two-dimensional finite element analysis of manufacturing processes, for three-dimensional analysis of manufacturing processes involving medium size models, and for preconditioning CG iterative solvers. In addition to what is mentioned earlier, it is worth noting that the solution accuracy of parallel direct solvers is identical to that of sequential direct solvers and that parallel direct solvers can also be utilized with solution decomposition approaches regardless of the problem to be solved.

Farhat and Wilson (1988) were among the first researchers to present parallel direct solvers. Parallelization can be applied for local memory processors as well as for shared memory processors, where the first is typically applied to a cluster of multiple computers, whereas the latter typically would be one computer with multiple threads.

Synn and Fulton (1995) were also among the first researchers to propose evaluation procedures to predict the performance of parallel direct solvers using skyline matrix storage. The usage of skyline or more efficient compressed sparse row storage formats is very often utilized in computer programs as it explores the large sparsity of the equation systems that are typical of finite element models in manufacturing.

In the present chapter a parallel direct solver for shared memory processors is presented and evaluated. The implementation is carried out in a personal computer (PC) due to increasing requirements from industry for running finite element simulations in standard hardware equipped with several cores and threads with shared memory. It is therefore an obvious request that the finite element computer programs can utilize multiple threads to reduce the computation time.

The direct solver to be presented is parallelized by OpenMP instructions in a FORTRAN implementation. The structure of the solver is explained in the chapter, and its performance is evaluated in terms of speed-up and efficiency. Amdahl's law (Hill and Marty, 2008) is applied to evaluate the amount of the code being parallel and to evaluate the potential benefit associated with an increasing number of threads. The entire source code is provided in the Appendix for the benefit of those who develop finite element computer programs. Only the call to the solver in existing computer programs using the same matrix storage format has to be replaced by a call to the presented solver. The requested inputs are the stiffness matrix in appropriate storage format together with the corresponding pointers to the diagonal positions, the right-hand side, the number of equations, and the number of threads to be utilized. A special implementation of the parallel direct solver that can be effectively and efficiently utilized as a preconditioner of parallel CG iterative solvers is also discussed.

A benchmark test case and a resistance spot welding test case selected from an industrial application enrich the overall presentation. The industrial example requires solution of a coupled electro-thermo-mechanical model in three dimensions with contact between individual hexahedral meshes. The case is further complicated by having large temperature gradients across the contacting interfaces, and therefore also large gradients in the mechanical stress response.

1.2 Brief review of standard gauss elimination

1.2.1 Skyline matrix storage

The presentation of the parallel direct solver is focused on the equation systems resulting from typical finite element assemblies,

$$[A]\{x\} = \{b\} \tag{1.1}$$

where $[A]$ is a symmetric, positive-definite, $n \times n$ matrix and $\{x\}$ and $\{b\}$ are $n \times 1$ vectors containing the unknowns and the right-hand side, respectively.

Due to symmetry of the system matrix, only half of the matrix needs to be built and stored (slightly more than half due to storage of all diagonal positions). Furthermore, since most finite element equation systems are sparse and by proper node numbering have many zeros far from the diagonal, it is common to employ skyline storage to reduce memory requirements by omitting all zeros above the highest skyline positions (Fig. 1.1). Zeros may still exist below this envelope as the skyline encloses all nonzero positions.

In skyline matrix storage, the equation system is typically stored in a one-dimensional vector $\{s\}$ with an additional index vector $\{i\}$ pointing to the diagonal positions. This is illustrated in Fig. 1.2 up to the 7th column. The size of the skyline vector will be the number of positions under the skyline. The size of the index vector equals the number of rows or columns n. Then, it follows that the size of the skyline vector is i_n as the last diagonal is the last position in the skyline vector.

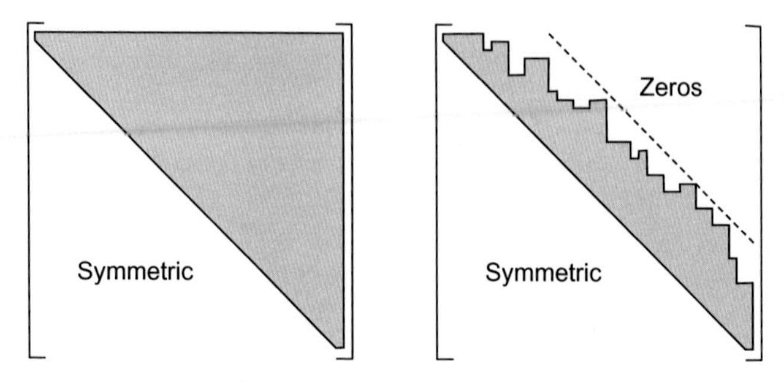

Fig. 1.1 System matrix storage by utilizing symmetry (left) and skyline format by omitting zeros (right). Only the grayed positions are stored. The dashed line covering the highest skyline positions indicates the requirements in case of banded matrix storage format.

Fig. 1.2 Format of skyline vector $\{s\}$ and index vector $\{i\}$ based on the original system matrix $[A]$. Numbers correspond to the position in the skyline vector.

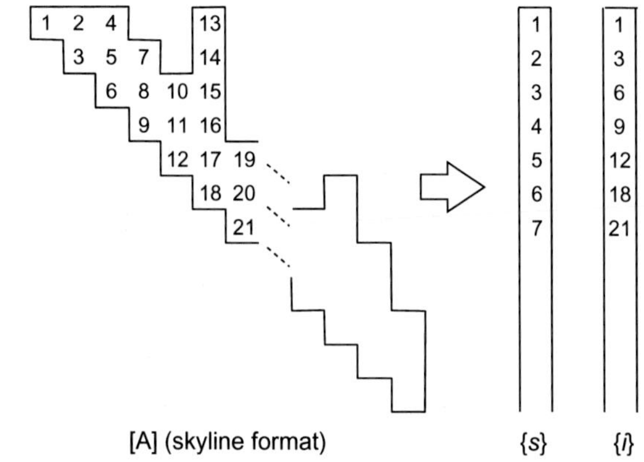

1.2.2 Gauss elimination

Among various solution techniques used by direct solvers, Gauss elimination was chosen due to its wide utilization and adequacy for parallelization. The solution of an equation system by Gauss elimination with column reduction comprises the following three basic steps:

- factorization of the system matrix and reduction of the right-hand side vector (this step is performed column by column, thereby being "with column reduction")
- division of the right-hand side vector by the diagonals of the system matrix
- backward substitution.

1.2.2.1 Factorization of the system matrix and reduction of the right-hand side vector

The first step in the factorization of the system matrix is illustrated by Fig. 1.3A for column j assuming that all columns $<j$ have been processed. The number of

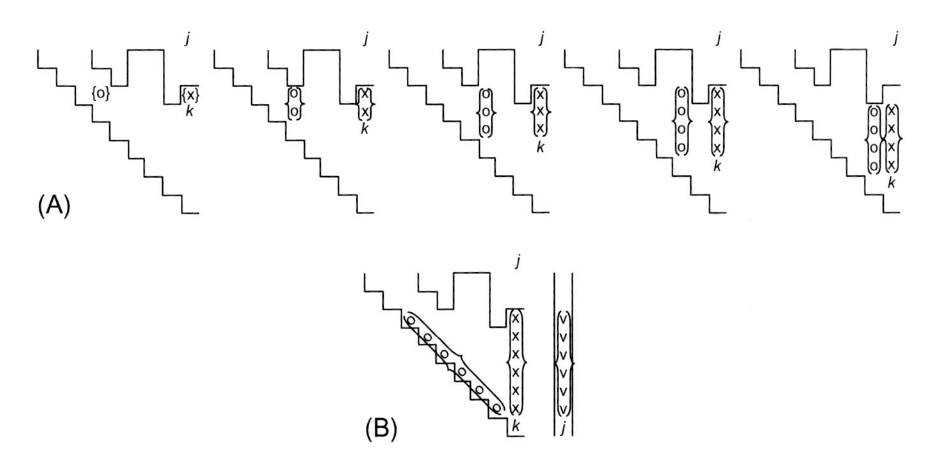

Fig. 1.3 Factorization of system matrix and reduction of right-hand side. (A) Reduction of off-diagonal positions in column j by subtraction of dot product formed by "o"s and "x"s. (B) Reduction of diagonal positions in system matrix and reduction of right-hand side vector.

operations equals the active column height −2 as the diagonal position and the topmost position are not processed. The illustration in Fig. 1.3A has active column height 7 and five operations are depicted in the subfigures. In each operation, the kth position is subtracted from the dot product formed by the vectors marked by "o"s and "x"s, i.e.

$$s_k = s_k - \{s_{\{o\}}\} \cdot \{s_{\{x\}}\} \tag{1.2}$$

Hereafter, all positions above the diagonal are divided by the diagonal position in the same row. In Fig. 1.3B, this corresponds to

$$\{s_{\{x\}}\} = \frac{\{s_{\{x\}}\}}{\{s_{\{o\}}\}} \tag{1.3}$$

where division is position-wise. This is followed by reduction of the diagonal term by

$$s_k = s_k - \{s_{\{x\}}\} \cdot \left\{s_{\{x\}}^{old}\right\} \tag{1.4}$$

which means subtraction of each multiplication of new and old off-diagonal position. New is defined as "after Eq. (1.3)" and old is defined as "before Eq. (1.3)."

Finally, the jth position in the right-hand side vector is reduced by subtraction of the dot product spanned by the marked "x"s and "v"s in Fig. 1.3B. Note that the positions marked by "x"s are now the latest updated, meaning after (1.4). The reduction of the right-hand side vector can be written as

$$b_j = b_j - \{s_{\{x\}}\} \cdot \{b_{\{v\}}\} \tag{1.5}$$

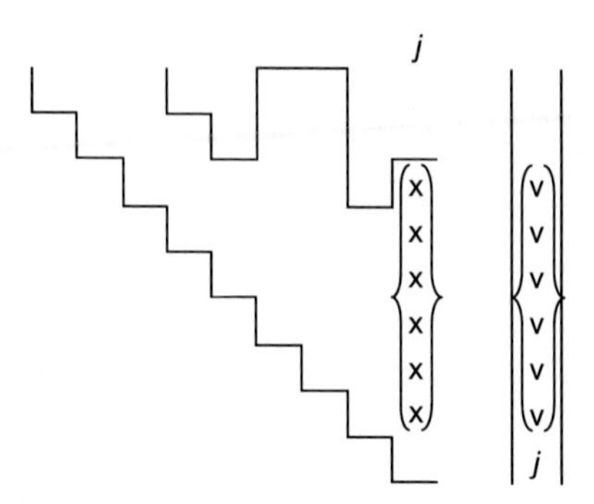

Fig. 1.4 Backward substitution by modification of positions $<j$ in right-hand side vector by subtraction of system matrix positions marked by “x”s multiplied by the *j*th position of the right-hand side vector.

1.2.2.2 Division of right-hand side vector by system matrix diagonals

This step is straight forward: Each position, j, in the right-hand side vector is divided by the system matrix diagonal term from the *j*th column. This completes the Gaussian elimination, such that all unknowns 1 to $j-1$ are eliminated from the *j*th row.

1.2.2.3 Backward substitution

The unknowns are now found by backward substitution and stored in the right-hand side vector $\{b\}$. The right-hand side vector is processed backward, such that Fig. 1.4 illustrates the substitution for the *j*th position assuming all positions $>j$ already processed. The positions marked by “v”s in the figure are modified by subtraction of positions in the system matrix marked by “x”s in the *j*th column. Before subtraction, these positions are multiplied by b_j, the *j*th position in the right-hand side vector, that is

$$\{b_{\{v\}}\} = \{b_{\{v\}}\} - b_j\{s_{\{x\}}\} \tag{1.6}$$

The right-hand side vector now includes the unknowns: $\{x\} = \{b\}$.

1.3 Structure of the parallel direct solver

The parallelization of the direct solver with skyline matrix storage is explained in this section with variable names matching the source code included in the Appendix. The factorization of the system matrix and reduction of the right-hand side (Section ***1.2.2.1***) is parallelized, whereas the division of the right-hand side vector by the system matrix diagonals (Section ***1.2.2.2***) as well as the backward substitution (Section ***1.2.2.3***) is left sequential as the time spent on these tasks is marginal compared to the factorization and reduction. The remaining of this section is devoted to the description of the parallelized part of the solver.

A few variables are introduced preliminary:

`j`	Column number
`jmax`	Latest processed column number
`kmax`	Latest processed diagonal position
`nthreads`	Number of threads
`ithread`	Actual thread number, ithread $\in \{0, 1, \ldots, \text{nthreads} - 1\}$
`iloop`	Counter for commenced column in each thread
`iquit`	Flag identifying when all columns have been processed
`skmatx`	Skyline vector $\{s\}$ containing the system matrix in skyline format
`maxa`	Index vector $\{i\}$ pointing to the diagonal positions in `skmatx`
`fmatx`	Right-hand side vector $\{b\}$

1.3.1 Parallel region

The factorization and reduction are performed within a parallel region defined by the following listing:

```
jmax=1
kmax=1
!$OMP parallel default (none) &
!$OMP private (...private variables...) &
!$OMP shared (...shared variables...)
  ithread=omp_get_thread_num()
  iloop=0
  iquit=0
  do while (iquit.eq.0)
    ! To be described
  enddo
!$OMP end parallel
```

The first diagonal position in the skyline, corresponding to the entire first column, does not change during the factorization. Therefore `jmax` and `kmax` are both initialized 1 (as if they were already processed).

Hereafter, the parallel region is defined starting from "`!\$OMP parallel`" and ending by "`!\$OMP end parallel`". In the beginning of the parallel region, all variables are identified as either private or shared. Typical private variables are actual thread number, counters, and intermediate results, which are unique variables on each thread. The shared group of variables consists of variables that are read or modified on all threads. In case of modifications, it is important to know, which thread is modifying, and then let the remaining threads wait if they are about to modify the same variable. Within the parallel region, each thread works independently, except for the shared variables. The first instructions are to get the actual thread number, `ithread`, and initialize `iloop` and `iquit`. This is followed by the main loop (`do while (iquit.eq.0)`) which will loop through all columns until the factorization and reduction have finished.

1.3.2 Main loop

Each thread is assigned a certain column to process in each cycle. The main loop takes the following form:

```
do while (iquit.eq.0)
   iloop=iloop+1
   j=(iloop-1)*nthreads+ithread+2
   if (j.gt.ntotv) then
      iquit=1
      exit
   endif
   jr=maxa(j-1)
   jd=maxa(j)
   jh=jd-jr
   is=j-jh+2
   ie0=0
!$OMP flush (kmax)
   do while (kmax.lt.jd)
      ! To be described
   enddo
enddo
```

The thread's counter, `iloop`, is incremented during each cycle. Based on the local counter and the total number of threads, each thread is assigned a column, `j`, to process. If there are no more columns to process, `iquit=1` to terminate the specific thread. This might happen while other threads still process their last column.

Fig. 1.5 illustrates the numbering of columns and the assignment of threads. As long as there are still columns to process, a few characteristic positions and dimensions are specified. The positions are the diagonal positions `jr` and `jd` of the `(j-1)`th and the `j`th column, respectively, and the dimension is the active height `jh` of column `j`. Finally, `is` is specified as the row number of the second active position in column `j`. These are all illustrated in Fig. 1.6. Before processing the column, a private variable, `ie0`, is initialized. This variable is later used as an indicator of the amount of the column to be processed without waiting. Variable `kmax` is flushed before the core loop, meaning that it is updated in all threads, such that a change in one thread will be visible for all other threads.

1.3.3 Core loop

The core loop is the processing of the `j`th column, and it is continued until the diagonal term has been processed (identified as `kmax=jd`), meaning that the entire column has been processed. The core loop is as follows:

```
do while (kmax.lt.jd)
   ! Initializations
   ihesitate=0
   ie0old=ie0
```

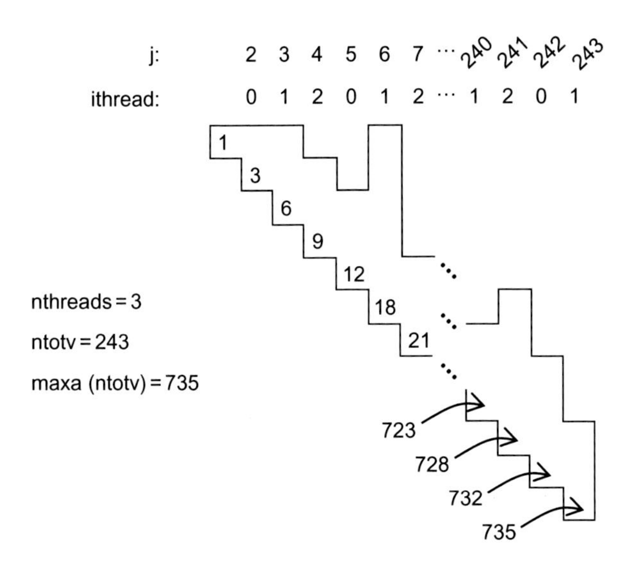

Fig. 1.5 Schematic skyline matrix with 243 degrees of freedom and a total skyline vector length 735. The example shows processing on three threads. The numbers in the diagonal correspond to the diagonal positions in the skyline vector. Numbers above the matrix show column number, `j`, and the thread number, `ithread`, of the thread to process a given column.

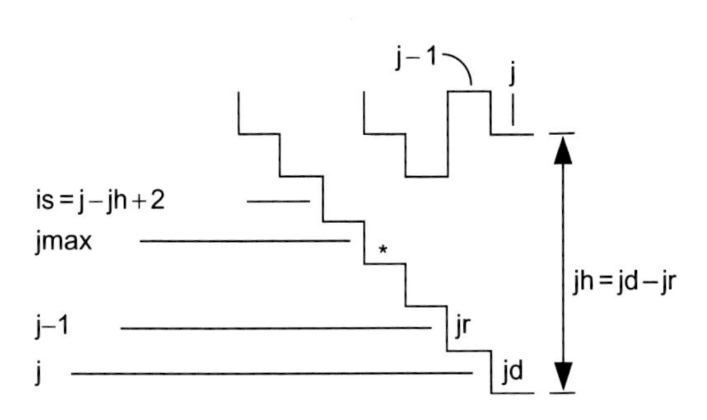

Fig. 1.6 Defined parameters for processing of column `j`. The example illustrates fully processed columns until the column with a "*" in the diagonal, which allows identification of `jmax` and hence also `kmax=maxa (jmax)`.

```
!$OMP flush (jmax,kmax)
   ! Judge if hesitation is necessary
   if (kmax.lt.jr) then
      ihesitate=1
      ie0=jmax
   endif
   if (jh.eq.2) then
      ! Reduce diagonal term
      ! Reduce right hand side
   elseif (jh.gt.2) then
      ! Reduce all equations except diagonal
      ! Reduce diagonal term
```

```
        ! Reduce right hand side
      endif
      if (ihesitate.eq.0) then
!$OMP critical
        if (j.gt.jmax) then
          jmax=j
          kmax=jd
        endif
!$OMP end critical
      endif
!$OMP flush (jmax,kmax)
    enddo
```

A variable, ihesitate, is introduced and initialized in each cycle. It is initialized as zero to indicate that the entire column is ready to be processed, which is only the case if column j-1 has already been processed. The variable ie0 is saved in ie0old before modifications, such that it is later available from both the current and the previous cycles. At this stage it is necessary to know how far the other threads are in their calculations, so kmax and jmax are flushed. An if-statement (kmax<jr) is judging if the entire column can be processed. If the statement is false (meaning kmax=jr), ihesitate=0 is kept and the entire column is processed in one cycle. Otherwise, if the statement is true, column j-1 has not been fully processed and it is necessary to hesitate, ihesitate=1. Hesitate does not mean wait, but means that only part of the column may be processed. The amount of the column that can be processed is defined by ie0=jmax. In the example of Fig. 1.6, only two positions in the column can be processed in this cycle of the loop; namely, from is to ie0. Then, in the next cycle (or a later cycle) kmax and jmax will have been increased on another thread, and the processing can be continued from ie0+1 to the diagonal position or to a limit provided by the new jmax.

After judgment of ihesitate and ie0, the factorization and reduction can take place. If the active part of column j is one (jh=1), only the diagonal exists, and no processing is needed. If the active height is two (jh=2), only the diagonal and one other position exist and is=j, so only reduction of the diagonal is necessary. For all other heights (jh>2), reduction takes place in the diagonal and positions above. These two cases, jh=2 and jh>2, are detailed subsequently in Sections ***1.3.3.1*** and ***1.3.3.2***.

This is followed by an if-statement to allow processing only when ihesitate=0. When this part of the code is reached without hesitation (ihesitate=0), the entire column has already been processed, and jmax and kmax have to be updated in order to let other threads proceed with the information that column jmax=j has been processed and the maximum processed diagonal position is kmax=jd. These variables are flushed immediately after the assignment. The assignment itself of the variables is enclosed in a critical region specifying that only one thread at a time can write to the variables. This is to avoid other threads to overtake and assign older values.

1.3.3.1 Reduction of off-diagonal terms, diagonal term, and right-hand side vector (case: jh>2)

The general case with `jh>2` is explained first. Hereafter, the case with `jh=2` is a special case. The following listing shows the reduction of off-diagonal positions, reduction of diagonal positions, and finally reduction of the right-hand side.

```
ie=jd-1+(ie0-j+1)*ihesitate
k00=jh-j-1+ie0old
if (k00.lt.0) k00=0
k0=0
do k=max0(jr+2,jd-j+ie0old+1),ie
  ir=maxa(is+k0+k00-1)
  id=maxa(is+k0+k00)
  ih1=min0(id-ir-1,1+k0+k00)
  if (ih1.gt.0) then
    ih2=min0(id-ir-j+(j-1-k0-k00)*ihesitate,2-j+k0+k00+(j-1-k0-k00)*
ihesitate)
    if (ih2.lt.1) ih2=1
    skmatx(k)=skmatx(k)-dot_product(skmatx(k-ih1:k-ih2),skmatx(id-
ih1:id-ih2))
  endif
  k0=k0+1
enddo
if (ihesitate.eq.0) then
  ir=jr+1
  ie=jd-1
  k=j-jd
  do i=ir,ie
    id=maxa(k+i)
    d=skmatx(i)
    skmatx(i)=d/skmatx(id)
    skmatx(jd)=skmatx(jd)-d*skmatx(i)
  enddo
  fmatx(j)=fmatx(j)-dot_product(skmatx(jr+1:jr+jh-1),fmatx(is-1:is
+jh-3))
endif
```

The reduction of off-diagonal positions takes place in the first loop in which the operations illustrated by Fig. 1.3A are performed. In case of hesitation, `ihesitate=1`, only part of the operations can be performed in the current cycle.

Fig. 1.7 illustrates such an example. Reading the figure from left to right, the first cycle starts, and `jmax` and `kmax` have been flushed. The "*" marks the latest processed diagonal position in the example. Then, since the `(j-1)`th column has not yet been processed, the cycle is performed with hesitation. In the example in Fig. 1.7, only

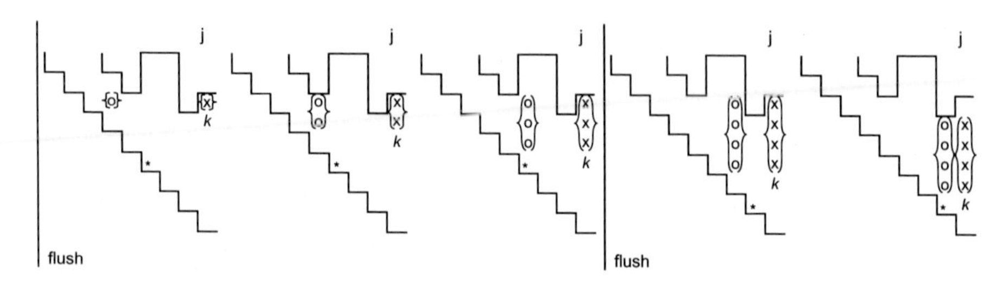

Fig. 1.7 Example of reduction of off-diagonal positions with hesitation. The "*" indicates latest processed diagonal position at latest flush.

the first three operations can be performed. Then a new cycle is started with flushing of `jmax` and `kmax`. Assume that the `(j-1)`th column has now been processed, as illustrated by "*." This cycle is thus performed without hesitation, and the remaining operations can be performed.

After the reduction of off-diagonal positions, the reduction of the diagonal position takes place inside the if-statement, which is only entered without hesitation, `ihesitate=0`. Immediately after, still only without hesitation, the right-hand side is reduced. These reductions follow the illustration and corresponding descriptions to Fig. 1.3B. The core loop (beginning of Section ***1.3.3***) will now reach the update of `kmax` and `jmax` without hesitation.

1.3.3.2 Reduction of diagonal term and right-hand side vector (case: jh=2)

The case where the active height of column `j` is equal to two is a special case of the above. There are no off-diagonals to process before reduction of the diagonal, so the algorithm is simplified to the following:

```
iwait=1
do while (iwait.eq.1)
!$OMP flush (jmax,kmax)
  if (kmax.ge.jr-1) then
    d=skmatx(jr+1)
    skmatx(jr+1)=skmatx(jr+1)/skmatx(jr)
    skmatx(jd)=skmatx(jd)-d*skmatx(jr+1)
    iwait=0
  endif
enddo
fmatx(j)=fmatx(j)-skmatx(jr+1)*fmatx(j-1)
ihesitate=0
```

A parameter, `iwait=1`, is defined and a loop will continue until this is set to `iwait=0`. The loop keeps flushing `kmax` until column `j-1` has been processed. Hereafter, the diagonal reduction and reduction of the right-hand side take place as mentioned earlier. The flags are set to `iwait=0` and `ihesitate=0` to proceed without hesitation to update `kmax` and `jmax` in the end of the core loop.

1.3.3.3 Source code

The direct solver with skyline storage is listed in the Appendix with the earlier mentioned parallel factorization of the system matrix and reduction of the right-hand side. The remaining tasks are sequential as the time spent on these is limited. These tasks are the division of the right-hand side vector by system matrix diagonals and the backward substitution.

1.3.4 Utilization as preconditioner of iterative solvers

In case of very large three-dimensional finite element models typical of manufacturing processes, CG iterative solvers may offer up to five times improvement in the overall CPU time against direct solvers. However, due to reasons previously mentioned in the introduction (Section **1.1**), there are several situations in which convergence of CG iterative solvers may result slower than that of direct solvers or simply does not exist. In these situations, it is advantageous to allow the computer program to automatically switch between direct and iterative solvers in order to always ensure convergence and to decide the best solution technique at every moment of simulation that is capable of optimizing the overall CPU time (Papadrakakis and Bitzarakis, 1996).

This justifies the utilization of a modified version of the proposed direct solver as preconditioner of CG iterative solvers. The modification involves adding a "drop off" instruction during reduction of the off-diagonal positions of the stiffness matrix as shown below for the general case with `jh>2`,

```
if (ihesitate.eq.0) then
  ir=jr+1
  ie=jd-1
  k=j-jd
  do i=ir,ie
    id=maxa(k+i)
    d=skmatx(i)
    skmatx(i)=d/skmatx(id)
    skmatx(jd)=skmatx(jd)-d*skmatx(i)
    ! Drop here if too small
    if (dabs(skmatx(i)).lt.drop) skmatx(i)=0.0d0
  enddo
endif
```

The "drop off" takes typical values in the order of `drop=1.0D-14` in order to reduce fill-in of the stiffness matrix by terms with very small values and, thereby, to reduce the total number of arithmetic operations utilized by the direct solver. This results in fast, approximate solutions, which may be used as preconditioning of CG iterative solvers.

1.4 Test cases and evaluation parameters

This section describes the benchmark test case utilized for evaluating the performance of different solvers and the welding test case chosen for evaluating the parallel direct solver in a real manufacturing application. The section also introduces definitions and parameters that will be utilized for assessing the overall efficiency of the parallelization procedure.

1.4.1 Benchmark test case

Fig. 1.8 shows the test case used in the comparison of different solvers implemented in IFORM3 (see Nielsen et al. (2013) for more information). The test is a simple upsetting of a cube between two flat parallel platens. Two of the cube faces have prescribed symmetry, and contact between the cube and the tools is frictionless. The cube with dimensions $10\text{mm} \times 10\text{mm} \times 10\text{mm}$ is compressed to half height through 100 simulation steps of $\Delta t = 0.05\text{s}$ with a velocity $v = 1\text{mm/s}$. The cube material is described by the flow stress curve $\sigma = 180.65\varepsilon^{0.183}\,\text{MPa}$. The cube is discretized by e^3 eight-node isoparametric elements of equal initial size. This discretization implies $(e+1)^3$ nodes and $n = 3(e+1)^3$ degrees of freedom with three unknown velocity components per node. The deformation of the cube is governed by the irreducible form of the finite element flow formulation

$$\delta\Pi = \int_V \overline{\sigma}\delta\dot{\overline{\varepsilon}}dV + K\int_V \dot{\varepsilon}_V\delta\dot{\varepsilon}_V dV \tag{1.7}$$

where V is the volume, δ denotes an arbitrary admissible variation in the velocity field, $\overline{\sigma}$ is the effective stress, and $\dot{\overline{\varepsilon}}$. is the equivalent strain rate. Volume constancy is enforced by penalizing the volumetric strain rate $\dot{\varepsilon}_V$ by a large positive number K.

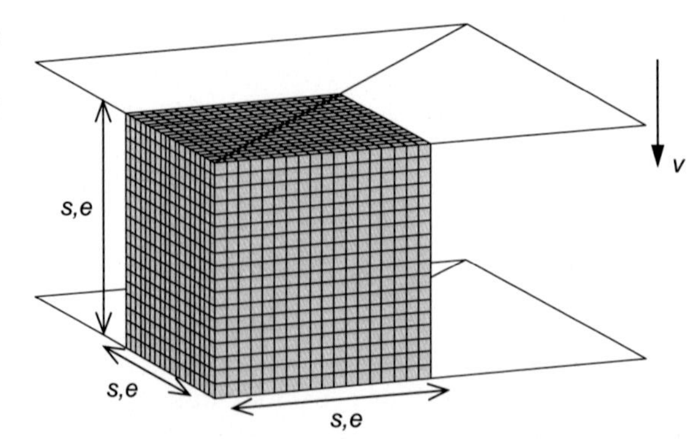

Fig. 1.8 Upsetting of cube with side length $s = 10\text{mm}$ discretized by e elements along each side. The lower platen is stationary, whereas the upper platen has velocity $v = 1\text{mm/s}$.

1.4.2 Resistance welding test case

The second evaluation of the parallel direct solver is a resistance welding case simulated in SORPAS 3D (see Nielsen et al. (2013) for more information). The case is simulated with different number of threads for evaluating the performance of the parallel direct solver. The welding case is shown in Fig. 1.9 and consists of two AISI 1008 steel alloy sheets of 1 mm thickness that are spot welded between two copper alloy electrodes with tip diameter 6 mm.

The electrode center axes are placed in a distance 13 mm to three of the sheet edges, but only 4 mm from the fourth edge. Total simulated process time is 340 ms. The electrode force is raised linearly to 3 kN within 20 ms and kept constant hereafter. AC current is applied after 40 ms, lasting 200 ms at a level of 8 kA RMS with a conduction angle of 80%. After the current is turned off, the electrodes keep the applied force for additional 100 ms to keep the sheets together while the weld solidifies.

The mesh shown in Fig. 1.9B consists of 7666 nodes, which in the mechanical model gives rise to 22,998 degrees of freedom, since there are three velocity components as unknowns in each node. In the electrical and thermal models, where the potential and the temperature are scalar fields, each node has one degree of freedom. Hence, the electrical and thermal models have 7666 degrees of freedom.

The mechanical model follows Eq. (1.7), and the electrical model is based on Laplace's equation solving the steady-state potential field. In the thermal model, conduction is modeled by a term similar to Laplace's equation, heat capacity is included in an additional transient term, and finally a term models the heat source due to the electrical current. For additional descriptions of the electrical and thermal models, refer to Zhang (2003). Contact conditions are included as well, see Zhang (2003) for electrical contact resistance, Nielsen et al. (2011) for mechanical contact, and Nielsen et al. (2013) for a comprehensive description of the computer implementation of the finite element flow formulation.

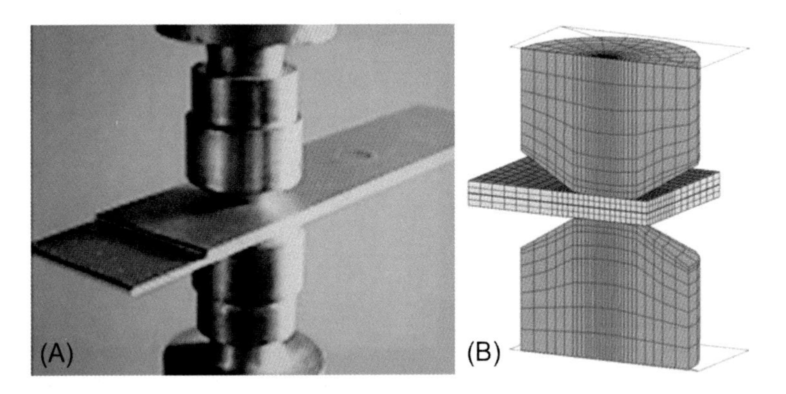

Fig. 1.9 Resistance spot welding test case. (A) Physical arrangement of electrodes and sheets for spot welding. (B) Applied mesh utilizing symmetry. Total number of nodes is 7666.

1.4.3 Evaluation parameters for parallelization

One metric commonly utilized to evaluate the performance of the parallelization is the speed-up defined as

$$\sigma = \frac{T_1}{T_N} \tag{1.8}$$

where T_1 is the solution time on one thread and T_N is the solution time using N threads. Ideally the speed-up would be $\sigma = N$. However, since only a fraction P of the program is parallel, the following inequality holds: $\sigma = \frac{T_1}{T_N} < N$. Efficiency is defined as the ratio of the obtained speed-up and the ideal speed-up, i.e.,

$$\eta = \frac{\sigma}{N} \tag{1.9}$$

The fraction of the program being parallel P influences directly the speed-up. The larger the fraction is the larger speed-up is achieved. Amdahl's (1967) law relates the estimated speed-up to this fraction as follows

$$\tilde{\sigma} = \frac{1}{(1-P) + \frac{P}{N}} \tag{1.10}$$

where $(1-P)$ represents the sequential part and $\frac{P}{N}$ represents the parallel part. From this equation it also follows that there is a limit of the obtainable speed-up. As the number of threads goes to infinity ($N \to \infty$), the sequential part still remains, and the limit becomes

$$\sigma_{\max} = \frac{1}{1-P} \tag{1.11}$$

Estimation of the fraction of the program being parallel for a given set of obtained speed-up and the number of threads is obtainable from rearrangement of Amdahl's law,

$$P = \frac{1 - \frac{1}{\sigma}}{1 - \frac{1}{N}} \tag{1.12}$$

1.5 Results and discussion

The first part of this section compares the performance of the proposed parallel direct solver with skyline matrix storage with that of a sequential CG iterative solver with preconditioning that was developed by Fernandes and Martins (2009). The second part analyses the performance of the parallel direct solver in a test case build upon the three-dimensional electro-thermo-mechanical finite element simulation of resistance welding.

All the simulations are performed on a laptop equipped with an Intel Core i7-3632QM processor with four cores and eight threads. It has 8 GB RAM and a clock frequency of 2.2 GHz. The operating system is Windows 10, 64-bit.

In order to keep the computer under the same global workload when testing the solution speed, all eight threads have been active during all simulations. When testing solution using N threads, the remaining $8-N$ threads have been running similar dummy simulations.

1.5.1 Solution time of different solvers

The benchmark test case described in Section ***1.4.1*** was solved with the two aforementioned different solvers for various mesh sizes. Fig. 1.10 shows the solution time, normalized with the solution time of the parallel direct solver using eight threads, as function of the number of degrees of freedom.

As expected from Amdahl's law, the solution time is not completely halved when going from sequential (one thread) to two threads, from two to four threads, and from four to eight threads because the computer program is not 100% parallel. This issue is dealt with in detail in Section ***1.5.3***.

The solution time in Fig. 1.10 for the sequential direct solver is produced by the parallel direct solver using one thread. It was tested that this solution time is practically identical to the original sequential solver, showing that the reorganizing of the code has not slowed down the code.

Comparing to the parallel direct solver using eight threads, the sequential iterative solver is slower below approximately 20,000 degrees of freedom, and above it is faster. When using fewer threads in the direct solver, this separation number of

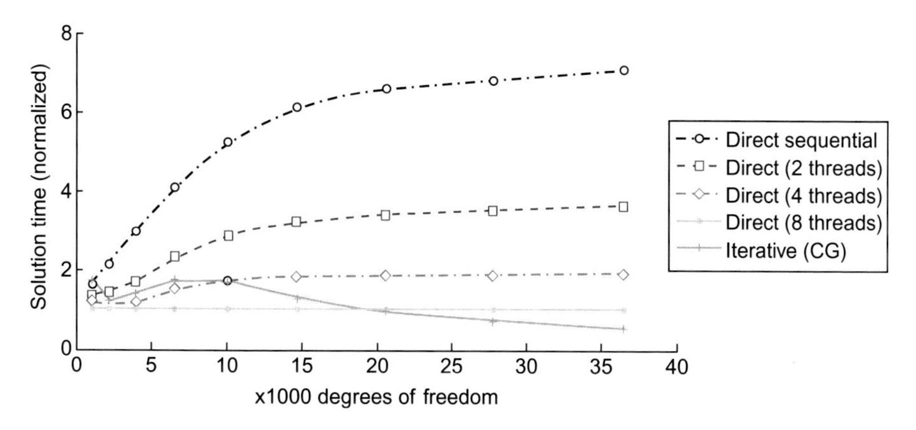

Fig. 1.10 Normalized solution time as function of degrees of freedom. The solution time is normalized by the solution time of the parallel direct solver using eight threads.

degrees of freedom is smaller. On the other hand, if more threads were available, the parallel direct solver would be faster than the sequential iterative solver at even larger number of degrees of freedom.

1.5.2 Accuracy of direct and iterative solvers

When the solution times of the iterative solver and the parallel direct solver are in the same range, the iterative solver has the benefit that other threads are still available for other purposes. However, it is worth mentioning that in contrast to the direct solver that is capable of providing significant reductions in the solution time as the number of threads increases (Fig. 1.10), the CG iterative solver is only capable of providing reductions of the solution time of approximately 25% as the number of threads increases. This conclusion in conjunction with the better overall accuracy of the direct solver justifies the reason why this chapter is focused on the parallelization of the direct solver instead of the CG iterative solver.

Fig. 1.11 shows an example of the accuracy differences between the two different solvers. The vertical stress component is shown on the cube after the final step, i.e., after compression to half height. The resulting stress distribution when applying the iterative solver varies as shown in Fig. 1.11 between −166.3MPa and −171.4MPa, whereas the distribution when applying the direct solver is uniform with a value of −168.8MPa. The deviations in the result from the iterative solver are −1.48% and

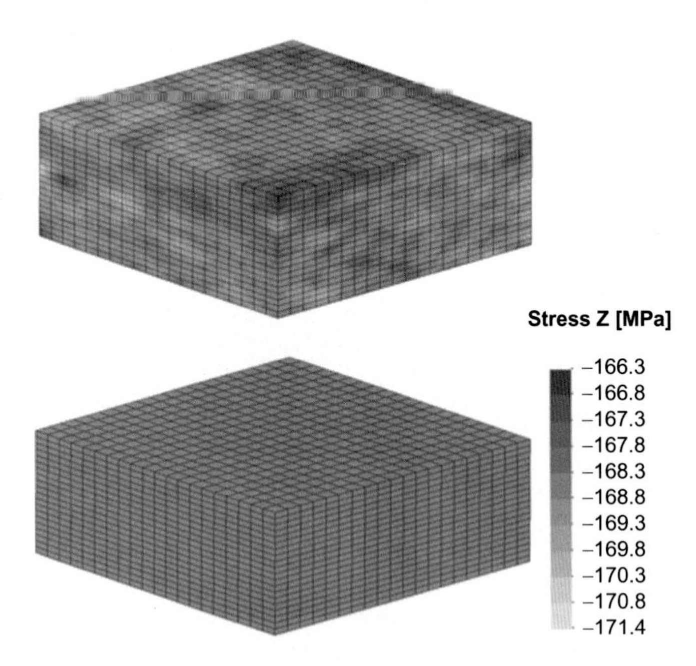

Fig. 1.11 Vertical component of the stress field in the cube compression example with 20 elements along each side by using the iterative solver (upper) and by using the direct solver with any number of threads (lower).

1.54% relative to the result from the direct solver. The aforementioned deviations could be reduced to −0.178% and 0.237% if the average number of iterations in the iterative solver was increased by 20%. However, this would increase the solution time by approximately 13% making it less attractive than it appears in Fig. 1.10.

The differences in accuracy shown in Fig. 1.11 can become significant when dealing with ill-conditioned systems of equations (Farhat and Wilson, 1988; Fernandes and Martins, 2009) resulting from rigid body motion and contact between deformable objects. This is the case in resistance spot welding because a large portion of the sheets undergo rigid body rotation and the contact pressure between the sheets, and the sheets and the electrodes, is crucial for properly estimating the electrical and thermal contact resistances. Small disturbances in the contact resistance will influence heat development, heat balance, and predicted geometry and size of the weld nugget. This explains the reason why the parallel direct solver is preferentially utilized for the analysis of resistance welding.

1.5.3 Performance of the parallel direct solver

Speed-up, efficiency, and the parallel fraction are evaluated based on the compression of a cube to half height in Section **1.4**. The speed-up (Eq. 1.8) is shown in Fig. 1.12 as function of degrees of freedom. The program is not entirely parallel, so the speed-up is less than ideal. Part of the program is still sequential, since only the equation solver of the main system of equations has been parallelized, and in addition heading (physical communication to and between the threads) takes time. As the system size (degrees of freedom) increases, relatively more time is necessary to solve the equation system, which means that the fraction of the code running in parallel becomes larger. This results in the larger speed-up seen in Fig. 1.12 at increasing number of degrees of freedom. It is also seen in the figure that the speed-up is largest for the smaller number of threads. This is a result of increased heading time and increased waiting time

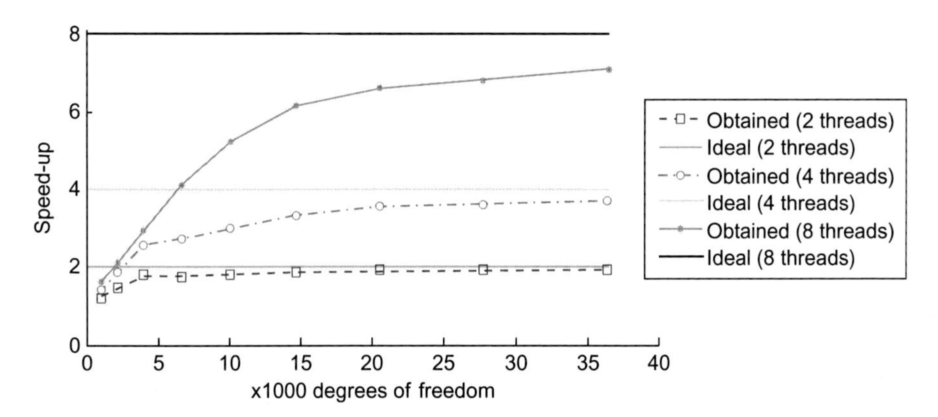

Fig. 1.12 Speed-up for 2, 4, and 8 threads as function of degrees of freedom.

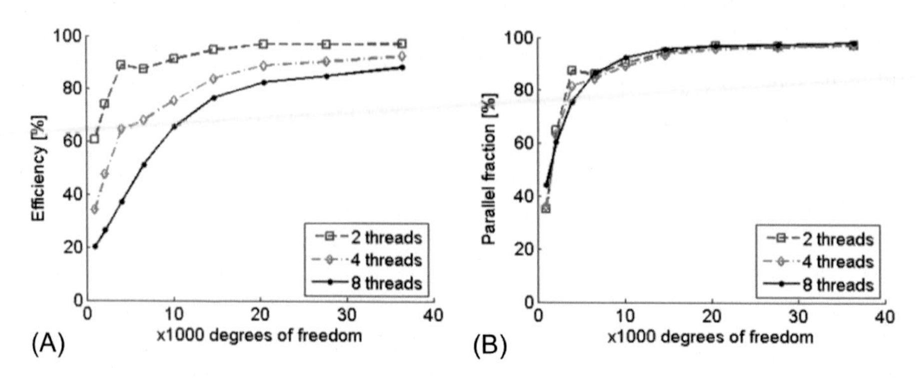

Fig. 1.13 Efficiency (A) and parallel fraction (B) for 2, 4, and 8 threads as function of degrees of freedom.

between threads when more threads are utilized, but it is also a result of a relatively smaller time fraction being parallel, simply because the solution time with more threads is shorter.

Similar conclusions can be drawn from Fig. 1.13A, which shows the efficiency (Eq. 1.9) as function of degrees of freedom for the different number of applied threads. Fig. 1.13B shows the parallel fraction (Eq. 1.12) which in agreement with the earlier discussion increases with system size. It has a steep increase in the beginning and then flattens out as the sequential part becomes small. In the end of the curve, the parallel fraction has reached 97%–98% for all number of threads. Note that this is of the entire program, not only the direct solver, which means that the CPU time taken in the rest of the program is negligible. Insertion into Eq. (1.11) shows that this corresponds to potential speed-up of 33–50 times if sufficient threads were available.

Assuming a parallel fraction of 97.5%, the potential speed-up is 40 and the speed-up as function of applied threads is given by Amdahl's law (Eq. 1.10). These are plotted in Fig. 1.14 together with the achieved speed-up using one (trivial), two, four,

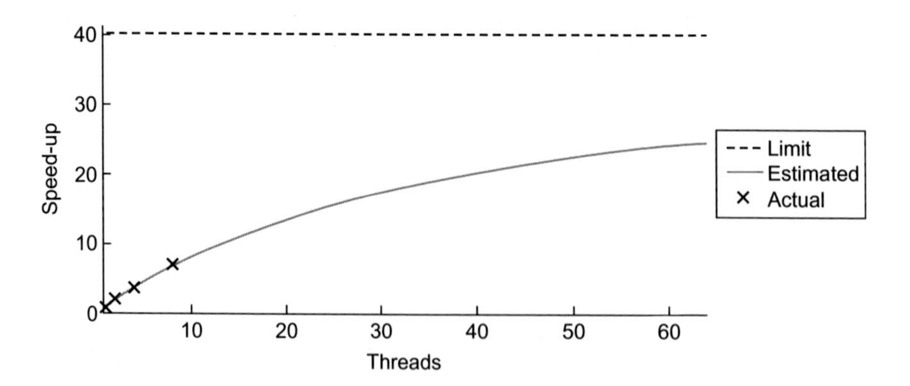

Fig. 1.14 Estimated speed-up as function of applied threads shown together with the theoretical limit and the actual speed-ups achieved for the cube example with 36,501 degrees of freedom.

and eight threads on the cube example with 36,501 degrees of freedom. This figure shows the potential of the parallel direct solver implementation as much more threads will be available on standard laptops in the future. The computer used in the present work has eight threads. This corresponds to an estimated speed-up of 6.8 (the achieved was 7.1).

In case 16 threads were available, the estimated speed-up is 11.6. Dreaming further to reach, e.g., 32 and 64 threads in standard laptops, the estimated speed-up is 18.0 and 24.9, respectively. Since the curve flattens out, it is also clear that if, e.g., two simulations are to be run, it is more efficient to run the two simulations simultaneously sharing the available threads rather than running one after the other using all threads.

1.5.4 Performance of the parallel direct solver in resistance welding

This subsection is built upon the resistance welding test case described in Section ***1.4.2***. Fig. 1.15 shows the resulting temperature field after the applied weld time. The spot seems almost axisymmetric showing that the chosen distance to the edge may not be a problem. However, this is without analysis of splash, which may be determining. Due to less material on the edge side, the temperature decreases slower near the edge, and this asymmetric cooling may result in a microstructure and residual stress distribution that the welding engineer has to be aware of.

The effect of node numbering optimization in the overall sparsity of the resulting finite element stiffness matrix is shown in Fig. 1.16, where the skyline height as function of the column number is shown before and after contact between the objects in Fig. 1.9B. The skyline heights are shown for three different cases. Fig. 1.16A and B shows the heights without node numbering optimization, Fig. 1.16C and D shows the resulting heights with node numbering optimization without information of contact between objects, and Fig. 1.16E and F shows the resulting heights with node numbering optimization with initial contact. The straight lines in the figures show

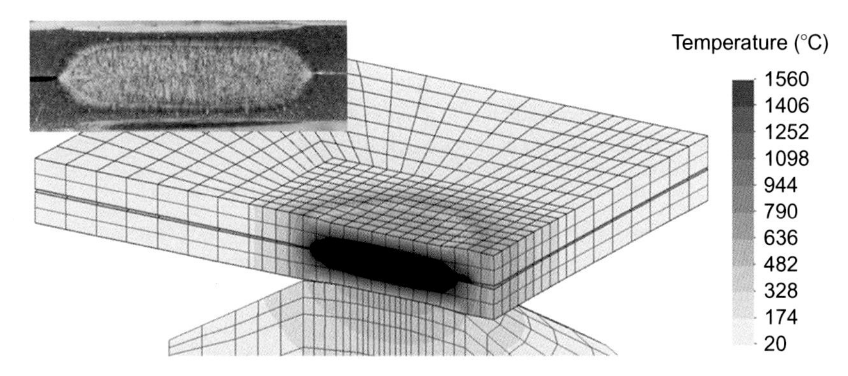

Fig. 1.15 Resistance spot welding test case (shown without upper electrode). The temperature field is shown after 240 ms, where the weld time ends. The inserted photo shows a weld nugget from a similar experimental study.

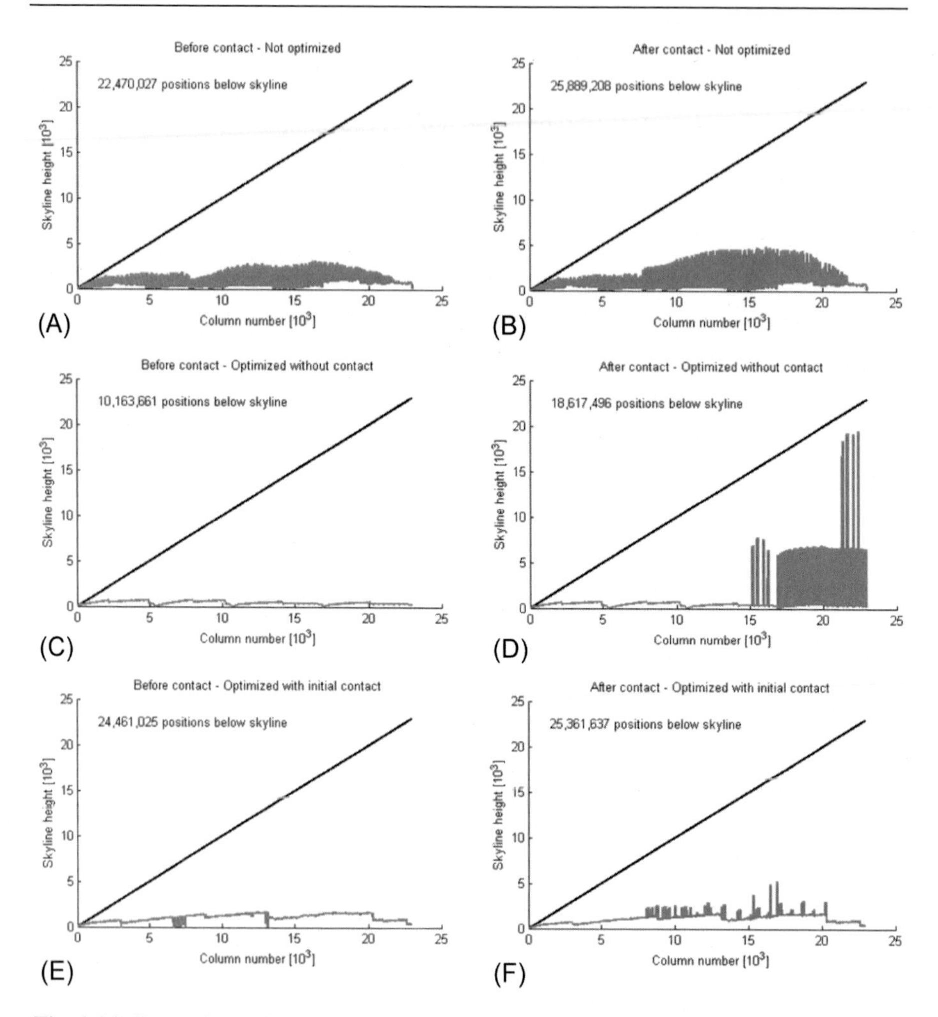

Fig. 1.16 Comparison of skyline topology before (A, C, E) and after (B, D, F) contact depending on optimization of node numbering. (A and B) Without optimization. (C and D) With optimization independent of contact. (D and E) With optimization including initial contact. The straight lines show the height corresponding to a full matrix (halved due to symmetry). Number of matrix positions below the skyline is noted in each figure.

the height corresponding to a full matrix, i.e., to a full upper triangular matrix due to symmetry. This shows, for all the cases in Fig. 1.16, the importance of an efficient storage format, as the skyline format applied here.

Without considering contact (optimization of node numbering before bringing the electrode meshes and sheet meshes in contact), the immediate benefit of node numbering optimization is seen by comparing Fig. 1.16A and C. Fig. 1.16A shows the skyline height of the original mesh, and Fig. 1.16C shows the skyline height of the optimized mesh, where the number of matrix positions below the skyline is more than halved. Of more interest for the calculation time is the skyline height after the objects have been brought into contact. When the optimization is made without considering contact, peaks will typically appear as in Fig. 1.16D. The differences in node numbers between the contacting nodes are large, and hence the skyline heights peak due to contact. Comparing Fig. 1.16D to B, it is seen that the number of matrix positions below the skyline is still reduced to about 72%.

Fig. 1.16E and F shows the resulting skyline profile when optimization of node numbering is made with inclusion of the initial contact information. The number of matrix positions is not improved as much as before (in fact the number increases before contact and only decreases slightly with contact). However, the peaks due to contact are reduced significantly, which will contribute to better speed-up in case of parallel solving of the equation system.

The solution times and obtained speed-ups in the welding case are shown in Fig. 1.17 for the two approaches to node numbering optimization. Solution times and speed-ups are shown as function of applied threads, where, as in the earlier analyses, when applying N threads, the remaining $8-N$ threads have been applied to a similar dummy simulation. The solution times in Fig. 1.17A and C are shown for the total running time of the entire simulation as well as for the pure solution time of the equation system in the mechanical model and in the electrical and thermal models. These pure solution times are accumulated over the entire simulation. It is clear from the figure that the equation solving in the mechanical model is the main contributor to the total solution time. The combined solution time in the electrical and thermal models is much less, partly due to fewer iterations and in particular due to the smaller system size (7666 degrees of freedom compared to 22,998 degrees of freedom in the mechanical model). The figures also include the remaining time spent in the simulation, i.e., the total time subtracted from the pure solution time in the main equation systems. Thus the remaining solution time is a sum of setting up the equation systems, searching for and evaluating contact, updating variables before time stepping, etc.

The overall solution time decreases with increasing number of applied threads. This is mainly accommodated by the shorter time spent in the pure solution of the mechanical equation system. The time spent in pure solution of the electrical and thermal equation systems decreases only little, and the time spent on remaining tasks should be unchanged, since it is not parallelized. An interesting difference is observed between the two approaches to the node numbering optimization. When the optimization is performed without information of the contact, the solution time does not decrease noticeably when applying more than three threads (Fig. 1.17A), and correspondingly the speed-up does not increase noticeably when applying more than three threads (Fig. 1.17B). On the other hand, the solution time decreases, and the speed-up increases remarkably over the whole range of applied threads when node numbering optimization includes information on initial contact. The reason for this difference is explained by peaks in the skyline height due to contact (see Fig. 1.16D and E). In the case where node numbering optimization is performed without initial contact, the peaks are high and separated by shorter columns. This results in waiting time in the threads processing the shorter columns, and thereby

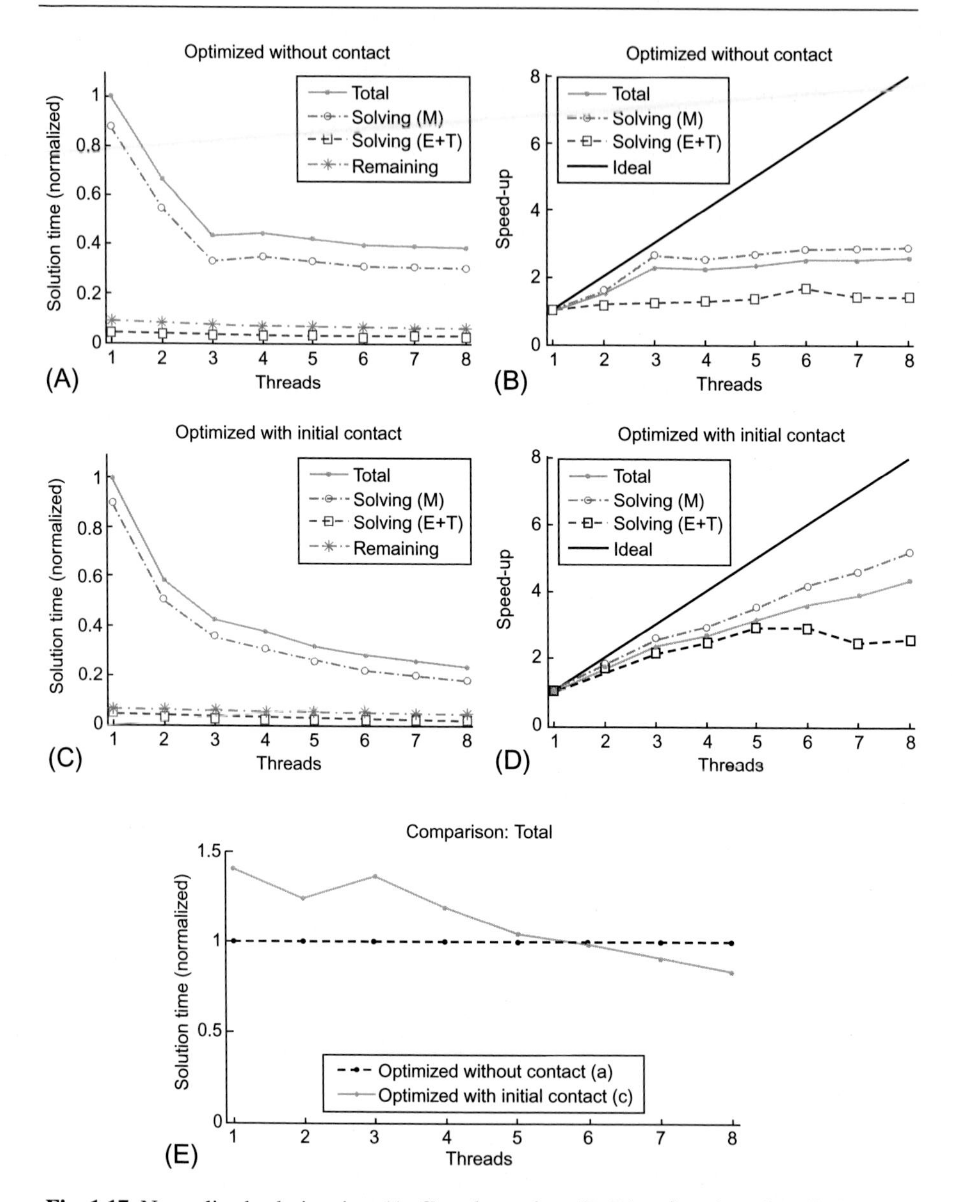

Fig. 1.17 Normalized solution time (A, C) and speed-up (B, D) as function of applied threads. (A and B) Optimization of node numbering independent of contact. (C and D) Optimization of node numbering with initial contact. (A–D) include the total solution time of the entire solution, the pure equation solving in the mechanical model (M), and the combined pure equation solving in the electrical and thermal models (E+T). The solution time of the “remaining” is also included, which is the total subtracted from the pure equation solving. A comparison of the total solution times is given in (E).

poorer speed-up. The column heights are more equal in case of optimization with initial contact, and therefore speed-up is better preserved.

In both cases, the speed-up of the pure solution time in the mechanical model is the highest, which is a result of the larger system size (22,998 degrees of freedom) compared to the electrical and thermal system sizes (7666 degrees of freedom). The speed-up of the total solution time is closer to that of the mechanical, since the mechanical model is the main contributor to the time. The total speed-up is slightly lower than the mechanical speed-up due to the lower speed-up in the electrical and thermal models and due to no speed-up in the remaining nonparallelized code.

When it comes to speed-up, the earlier comparison shows that the node numbering optimization taking initial contact into account is clearly better than the optimization without contact information. Fig. 1.17E compares the actual solution time of the two approaches for different numbers of applied threads. The solution times are normalized by the solution time of the optimization without contact information. The figure shows that the approach including initial contact (which has the better speed-up) is slower by a factor of 1.4 when using one thread. However, due to the better speed-up, it becomes faster when applying six or more threads. The reason for the slower solution when using few threads is that the initial contact is much more than the contact after separation of the sheets outside the weld zone, and therefore the optimized skyline according to the initial contact is not optimal throughout the entire solution.

Fig. 1.17C and D, as well as the figures related to the cube compression example, proves the presented parallel direct solver. Hereafter, it is up to a correct approach for the node numbering optimization to get the best use of it. In the specific welding case, an improved strategy would be to start out with an optimized node numbering based on the initial contact, and then reoptimize the node numbering when the sheets have separated outside the weld zone. This is implemented in SORPAS 3D by reoptimizing every time the number of contact pairs has changed by a certain amount.

1.6 Conclusions

With a sequential direct solver based on skyline matrix storage and using Gaussian elimination with column reduction as starting point, a parallel version is developed and tested. The sequential version is first explained for providing an overview. Second, the parallel implementation is discussed. An overview of the structure is given, and important sections of the source code are explained. The entire source code of the parallel direct solver is provided in Appendix.

The proposed parallel direct solver is compared with a sequential iterative solver using the conjugate gradient method with preconditioning. The iterative solver is the fastest solver, except in the interval up until around 20,000 degrees of freedom, where the proposed parallel direct solver with skyline matrix storage using eight threads is fastest. If more threads were available the parallel direct solver would also be competitive for larger system sizes.

Speed-up and efficiency have been evaluated in an upsetting example, showing a good parallelization. For the entire program, i.e., including the direct solver as well as the remaining of the program, Amdahl's law shows that the code is around 97.5% parallelized when dealing with the large system sizes. This corresponds to a maximum speed-up of 40, which is reached asymptotically with increasing number of threads applied.

The iterative solver is fast, but accuracy problems derived from ill-conditioned systems of equations resulting from rigid body motion and contact between deformable objects, among other problems, justify the utilization of parallel direct solvers for the electro-thermo-mechanical finite element analysis of resistance welding. The test case in resistance welding also justifies the importance of optimal node numbering. Depending on the approach for node numbering optimization, speed-up can be preserved over the entire range of tested number of threads.

The Appendix provides the source code of a parallel direct solver based on skyline matrix storage, which can be directly implemented in existing finite element codes. Hopefully, this will result in reduced solution time among researchers and engineers, and it is incited that improvements and further developments will be published for the common benefit.

Appendix

```
subroutine skyline_gauss_omp (skmatx,fmatx,&
                          maxa,nthreads,ntotv)
 use omp_lib
 implicit none
 !-----------------------------------
 ! This subroutine solves a regular system:
 !    skmatx*x=fmatx
 ! where
 ! skmatx  is the skyline vector of the system
 !         matrix,
 ! fmatx   is the right hand side vector (in)
 !         and later the vector of unknowns (out),
 ! x       is the vector of unknowns outputted
 !         through fmatx
 ! The index vector pointing to the diagonal
 ! positions in skmatx is maxa.
 ! The number of degrees of freedom is ntotv.
 ! Number of threads to use during solving is
 ! nthreads.
 ! Method: Gaussian elimination with column
 !         reduction.
 !
 ! This skyline solver was originally provided
 ! in sequential form by
```

```
    !   J.E.Akin, Finite Elements for Analysis
    !   and Design, Academic Press, London, 1993.
    ! It is parallelized in the present work by
    !   C.V.Nielsen and P.A.F.Martins
    !-------------------------------------
    integer i,id,ie,ie0,ie0old,ih1,ih2,ihesitate,&
            iloop,iquit,ir,is,ithread,iwait,j,jd,&
            jh,jmax,jr,k,k0,k00,kmax,nthreads,ntotv
    integer maxa(*)
    double precision d,fmatx(*),skmatx(*)
    ! Set number of threads
    call omp_set_num_threads (nthreads)
    ! Get (actual) number of threads
    nthreads=omp_get_max_threads()
    ! Initializations
    jmax=1
    kmax=1
  !$OMP parallel default (none) &
  !$OMP private (d,i,id,ie,ie0,ie0old,ih1,ih2,ihesitate,&
  !$OMP          iloop,iquit,ir,is,ithread,iwait,j,jd,jh,&
  !$OMP          jr,k,k0,k00) &
  !$OMP shared (fmatx,jmax,kmax,maxa,nthreads,ntotv,&
  !$OMP         skmatx)
    ! Factorize skmatx and reduce fmatx
    ithread=omp_get_thread_num()
    iloop=0
    iquit=0
    do while (iquit.eq.0)
      iloop=iloop+1
      j=(iloop-1)*nthreads+ithread+2
      if (j.gt.ntotv) then
        iquit=1
        exit
      endif
      ! Characteristic positions in skyline
      jr=maxa(j-1)
      jd=maxa(j)
      jh=jd-jr
      is=j-jh+2
      ! Start of core code
      ie0=0
  !$OMP flush (kmax)
      do while (kmax.lt.jd)
        ! Initializations
        ihesitate=0
        ie0old=ie0
```

```
!$OMP flush (jmax,kmax)
        ! Judge if hesitation is necessary
        if (kmax.lt.jr) then
           ihesitate=1
           ie0=jmax
        endif
        if (jh.eq.2) then
           ! Reduce diagonal term
           iwait=1
           do while (iwait.eq.1)
!$OMP flush (kmax)
              if (kmax.ge.jr-1) then
                 =skmatx(jr+1)
                 skmatx(jr+1)=d/skmatx(jr)
                 skmatx(jd)=skmatx(jd)-d*skmatx(jr+1)
                 iwait=0
              endif
           enddo
           ! Reduce right hand side (fmatx)
           fmatx(j)=fmatx(j)-skmatx(jr+1)*fmatx(j-1)
           ihesitate=0
        elseif (jh.gt.2) then
           ! Reduce all equations except diagonal
           ie=jd-1+(ie0-j+1)*ihesitate
           k00=jh-j-1+ie0old
           if (k00.lt.0) k00=0
           k0=0
           do k=max0(jr+2,jd-j+ie0old+1),ie
              ir=maxa(is+k0+k00-1)
              id=maxa(is+k0+k00)
              ih1=min0(id-ir-1,1+k0+k00)
              if (ih1.gt.0) then
                 ih2=min0(id-ir-j+(j-1-k0-k00)&
                   *ihesitate,&
                   2-j+k0+k00+(j-1-k0-k00)&
                   *ihesitate)
                 if (ih2.lt.1) ih2=1
                 skmatx(k)=skmatx(k)&
                       -dot_product(&
                       skmatx(k-ih1:k-ih2),&
                       skmatx(id-ih1:id-ih2))
              endif
              k0=k0+1
           enddo
           if (ihesitate.eq.0) then
              ! Reduce diagonal term
```

```
            ir=jr+1
            ie=jd-1
            k=j-jd
            do i=ir,ie
              id=maxa(k+i)
              d=skmatx(i)
              skmatx(i)=d/skmatx(id)
              skmatx(jd)=skmatx(jd)-d*skmatx(i)
            enddo
            ! Reduce right hand side (fmatx)
            fmatx(j)=fmatx(j)&
              -dot_product(skmatx(jr+1:jr+jh-1),&
                                  fmatx(is-1:is+jh-3))
          endif
        endif
        if (ihesitate.eq.0) then
!$OMP critical
          if (j.gt.jmax) then
            jmax=j
            kmax=jd
          endif
!$OMP end critical
        endif
!$OMP flush (jmax,kmax)
      enddo
    enddo
!$OMP end parallel
    ! Divide by diagonal pivots
    do i=1,ntotv
      id=maxa(i)
      fmatx(i)=fmatx(i)/skmatx(id)
    enddo
    ! Back substitution
     jd=maxa(ntotv)
     do j=ntotv,2,-1
       d=fmatx(j)
       jr=maxa(j-1)
       if (jd-jr.gt.1) then
        is=j-jd+jr+1
        k=jr-is+1
        fmatx(is:j-1)=fmatx(is:j-1)-skmatx(is+k:j-1+k)*d
       endif
       jd=jr
     enddo
    endsubroutine skyline_gauss_omp
```

Acknowledgments

The authors would like to acknowledge the support provided by Fundação para a Ciência e a Tecnologia of Portugal and IDMEC under LAETA—UID/EMS/50022/2013 and PDTC/EMS-TEC/0626/2014.

References

Al-Nasra, M., Nguyen, D.T., 1992. An algorithm for domain decomposition in finite element analysis. Comput. Struct. 39, 277–289.

Amdahl, G.M., 1967. Validity of the single processor approach to achieving large scale computing capabilities. In: AFIPS Conference Proceedings, pp. 483–485.

Farhat, C., 1988. A simple and efficient automatic FEM domain decomposer. Comput. Struct. 28, 579–602.

Farhat, C., Wilson, E., 1988. A parallel active column equation solver. Comput. Struct. 28, 289–304.

Fernandes, J.L.M., Martins, P.A.F., 2009. Robust and effective numerical strategies for the simulation of metal forming processes. J. Manuf. Technol. Res. 1, 21–36.

Hestenes, M.R., Stiefel, E., 1952. Methods of conjugate gradients for solving linear systems. J. Res. Natl. Bur. Stand. 49, 409–436.

Hill, M.D., Marty, M.R., 2008. Amdahl's in the multicore era. IEEE Comput. Soc. 41, 33–38.

Kim, S.Y., Im, Y.T., 2003. Parallel processing of 3D rigid-viscoplastic finite element analysis using domain decomposition and modified block Jacobi preconditioning technique. J. Mater. Process. Technol. 134, 254–264.

Lanczos, C., 1952. Solution of systems of linear equations by minimized iterations. J. Res. Natl. Bur. Stand. 49, 33–53.

Meijerink, J.A., van der Vorst, H.A., 1977. An iterative solution method for linear systems of which the coefficient matrix is a symmetric M-matrix. Math. Comput. 31, 148–162.

Nielsen, C.V., Martins, P.A.F., Zhang, W., Bay, N., 2011. Mechanical contact experiments and simulations. In: Steel Research International, Special Edition: 10th International Conference Technology of Plasticity, pp. 645–650.

Nielsen, C.V., Zhang, W., Alves, L.M., Bay, N., Martins, P.A.F., 2013. Modelling of Thermo-Electro-Mechanical Manufacturing Processes. Springer, London, UK.

Papadrakakis, M., Bitzarakis, S., 1996. Domain decomposition PCG methods for serial and parallel processing. Adv. Eng. Softw. 25, 291–307.

Synn, S.Y., Fulton, R.E., 1995. The performance prediction of a parallel skyline solver and its implementation for large scale structure analysis. Comput. Syst. Eng. 6, 275–284.

Zhang, W., 2003. Design and implementation of software for resistance welding process simulations. Trans. J. Mater. Manuf. 112, 556–564.

Optimal inspection/actuator placement for robust dimensional compensation in multistage manufacturing processes

2

J.V. Abellán-Nebot, I. Peñarrocha*, E. Sales-Setién*, J. Liu*[†]
*Universitat Jaume I, Castelló de la Plana, Spain, [†]The University of Arizona, Tucson, AZ, United States

Nowadays, industry is moving toward the implementation of manufacturing systems capable of adapting themselves in real time when changes or deviations in the production line arise. For this purpose, process knowledge should be applied for production performance estimation together with optimal control systems in order to keep the production quality within given specifications. In this chapter, a brief review of a modeling procedure of multistage manufacturing systems, named state-space model, is provided. The definition of the state-space model can then be used for ensuring process quality by the implementation of a feed-forward control strategy. Under this strategy, it is addressed the challenge of where to place the inspection units in order to estimate the state of the manufacturing system and where to place the controllers in order to enhance the process performance. The general procedure to deal with this problem is presented based on the explicit formulation of the predictive control problem subjected to the removal of the actuator and the inspection stations. A two-dimensional (2D) case study from sheet metal working industry is shown to illustrate the methodology.

2.1 Introduction

One of the most important challenges in modern industry is the implementation of manufacturing systems that are capable of producing "zero-defect" products. Many technological roadmaps from international research organizations, such as the Global Research of Business Innovation Program in Intelligent Manufacturing Systems (IMS2020, 2010) and the Research and Development policies from the European Commission (EFFRA, 2013), are promoting research activities toward the development of such advanced manufacturing systems. Two of the key research directions presented in these roadmaps are (i) the development of reliable process knowledge that accurately defines process behavior according to process parameters, control parameters, and process perturbations and (ii) the development and implementation of manufacturing systems that are adaptive in real time to keep final product quality within specifications.

Computational Methods and Production Engineering. http://dx.doi.org/10.1016/B978-0-85709-481-0.00002-1

Current state of industrial practice in process monitoring and fault detection is mostly based on statistical approaches, such as statistical process control (SPC) (Montgomery, 2007; Shi and Zhou, 2009). These techniques can rapidly detect and analyze significant variations along the manufacturing process. However, the identification of the root causes has been challenging for both researchers and practitioners (Liu, 2010). More advanced techniques require an explicit model of the manufacturing processes in order to detect significant changes in the production system and, furthermore, to identify their sources and actuate to mitigate or eliminate their effects. For this purpose, a key issue is the integration of manufacturing process knowledge through a mathematical model.

A variety of research works have been conducted for integrating process knowledge and improving process control to ensure product quality. Some of these works have been focused on the implementation of artificial intelligent models, such as artificial neural networks or fuzzy logic systems (Niaki and Davoodi, 2009; Zhang and Jiang, 2011). In other works, the process knowledge integration has been conducted through statistical analysis of experimental data such as autoregressive models (Paik, 2008; Pan et al., 2016). In all these methods, it is required experimental data or domain expertise to model process behavior. However, a more proper approach to extracting the real process behavior of a manufacturing system without experimental data is the use of engineering-driven models based on mechanistic/kinematic relationships (Shi and Zhou, 2009). This kind of models can be applied in multistage manufacturing systems, such as in assembly processes or machining processes, where clear mechanistic/kinematics relationships are available. In this field, one of the most successful approaches applied recently by the researchers is the stream-of-variation (SoV) model (Shi, 2006), a model adapted from the state-space model in control theory for modeling multistage manufacturing systems. The SoV model can mathematically define the kinematic relationships within stages and across stages in a multistage manufacturing process to evaluate and predict the performance of the entire system.

This chapter briefly reviews the state-space model approach for modeling multistage manufacturing systems, with special emphasis given to the model uncertainty in order to completely define the behavior of the process. It should be remarked that explicitly considering model uncertainty is critical when applying the presented approach in real-world projects, where a large number of stages and process parameters are expected. If omitted, the unrealistic nominal state-space model may lead to misleading results.

Given the state-space model, a feed-forward control strategy (Izquierdo et al., 2007) can be adopted in order to compensate upstream deviations and keep product quality variation within limits. In multistage manufacturing processes with a large number of stations, there are no straightforward solutions to the questions such as where to place in-process measurement stations to estimate current process state and where to place actuation stations to adapt their parameters according to previous estimations. This chapter presents these questions and proposes a two-step methodology to optimally place both actuator and inspection stations.

2.2 State-space approach for modeling multistage assembly processes

2.2.1 Stage-level model

Among multistage manufacturing processes, one of the most common processes is the assembly of sheet metal panels by a series of welding operations at different stations. In this type of multistage assembly processes (MAPs), a four-way hole and a two-way slot are usually used to locate a part or a subassembly. The center of a four-way hole and the center of a two-way slot are chosen as the reference points, which are used to define a part coordinate system in a two-dimensional (2D) space, as shown in Fig. 2.1. These two points are denoted as $LP_{1,i,k}$ and $LP_{2,i,k}$ for a part, and as $LS_{1,s,k}$ and $LS_{2,s,k}$ for a subassembly, denoting the four-way centers of the ith and the sth parts/subassembly at stage k, respectively. The position of part i at stage k can then be represented by its position of $LP_{1,i,k}$, i.e., ($LP_{1,i,k}(x)$, $LP_{1,i,k}(z)$), and an angle, $\beta_{i,k}$, between the line connecting $LP_{1,i,k}$ and $LP_{2,i,k}$ and the X axis, as shown in Fig. 2.1.

At each station, a fixture based on a four-way pin and a two-way pin, also named as concentric and radial locators, is equipped to fix the position and the orientation of a part/subassembly (Fig. 2.1). These locators may not be perfectly placed in the fixture or may be loosen or worn out during their normal operation. Such imperfections make the sheet metal parts/subassemblies to randomly deviate from their nominal positions and/or orientation. This position and orientation deviations can be modeled as a function of the deviation of the concentric and radial locators with a kinematic model.

For modeling purposes, two types of coordinate systems (CSs) are used, namely, global CS and part CS. The global CS remains unchanged for all stages, and a part CS is attached to a particular part/subassembly. Each part/subassembly is characterized by its deviation from the nominal position. In sheet metal assembly processes, the deviations can be represented by two translational coordinates, X and Z, and one rotational coordinate, β.

Fig. 2.2 shows how a four-way locator deviation in Z direction will alter the position and orientation of a part. As derived in Shi (2006), the mathematical relationship

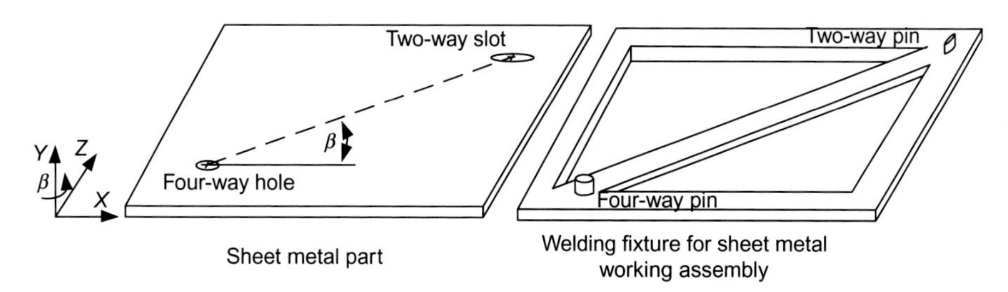

Fig. 2.1 Sheet metal part with a four-way hole and a two-way slot for placement and common fixture used in sheet metal assembly processes based on four-way and two-way pins.

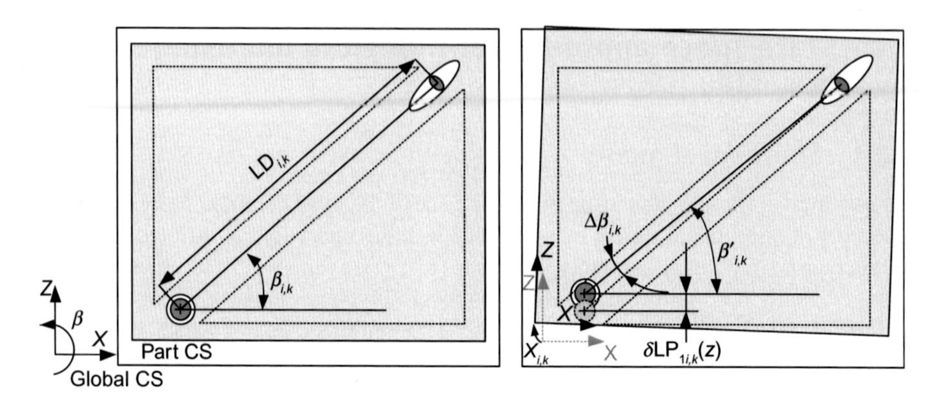

Fig. 2.2 Sheet metal part deviation due to a four-way locator deviation in *Z* direction.

between all potential locator errors (deviations in *X* and *Z* of the 4-way pin and deviation in *Z* of the 2-way pin) and part position and orientation deviations is

$$\boldsymbol{x}_{i,k}=\begin{bmatrix}\delta x_{i,k}\\ \delta z_{i,k}\\ \delta\beta_{i,k}\end{bmatrix}=\boldsymbol{R}_1^{i,k}\cdot\begin{bmatrix}\delta\mathrm{LP}_{1,i,k}(x)\\ \delta\mathrm{LP}_{1,i,k}(z)\\ \delta\mathrm{LP}_{2,i,k}(x)\\ \delta\mathrm{LP}_{2,i,k}(z)\end{bmatrix}\quad \boldsymbol{R}_1^{i,k}=\begin{bmatrix}1 & 0 & 0 & 0\\ 0 & 1 & 0 & 0\\ \dfrac{\sin\beta_{i,k}}{\mathrm{LD}_{i,k}} & \dfrac{\cos\beta_{i,k}}{\mathrm{LD}_{i,k}} & -\dfrac{\sin\beta_{i,k}}{\mathrm{LD}_{i,k}} & \dfrac{\cos\beta_{i,k}}{\mathrm{LD}_{i,k}}\end{bmatrix} \tag{2.1}$$

where $\mathrm{LD}_{i,k}$ is the distance between $\mathrm{LP}_{1,i,k}$ and $\mathrm{LP}_{2,i,k}$ and $\boldsymbol{x}_{i,k}$ denotes the deviation of part *i* with respect to the three coordinates (*X*, *Z*, and β) at stage *k*. Note that instead of directly using the global CS, the actual deviations of a part/subassembly from their nominal positions in space are defined with respect to the part CS using the vector $\boldsymbol{x}_{i,k}$.

2.2.2 System-level model

As shown in Fig. 2.3, consider two consecutive intermediate stations in a MAP, where two parts are fixed by their corresponding locators at station *k*. Assume that one of these locators is deviated, and the welding process generates a subassembly from the two parts. This subassembly is then transferred to Station $k+1$ and is fixed by a 4-way and a 2-way locator. It can be seen that the deviation generated at Station *k* is then propagated to station $k+1$, making the whole three-part assembly deviate from its nominal values even though all the locators at Station $k+1$ are located as designated, as shown in Fig. 2.3.

Mathematically, the propagation of random deviations can be modeled with a reorientation matrix, which represents the assembly process that transfers and reorientates the subassembly from an upstream station to an immediate downstream station. As shown in (Shi, 2006), the deviations of a subassembly in current station when it is moved from a previous station and mounted on a perfect fixture (error free) can be represented by

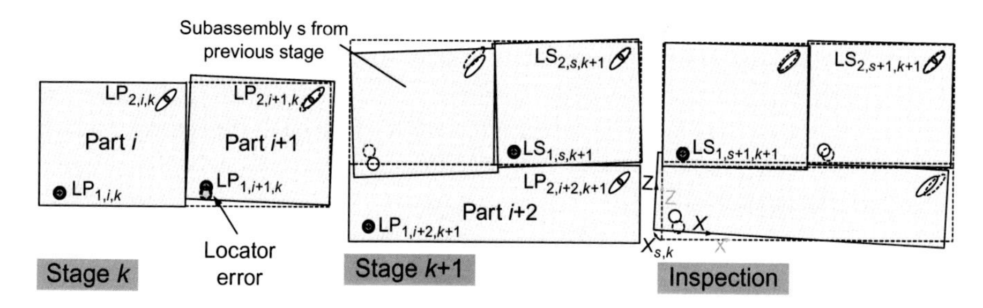

Fig. 2.3 Illustration of deviation propagation in multistage assembly processes.

$$\boldsymbol{x}_{s,k} = \begin{bmatrix} \delta x_{s,k} \\ \delta z_{s,k} \\ \delta \beta_{s,k} \end{bmatrix} = \boldsymbol{R}_2^{s,k-1} \cdot \begin{bmatrix} \delta \mathrm{LS}_{1,s,k-1}(x) \\ \delta \mathrm{LS}_{1,s,k-1}(z) \\ \delta \mathrm{LS}_{2,s,k-1}(x) \\ \delta \mathrm{LS}_{2,s,k-1}(z) \end{bmatrix} \quad \boldsymbol{R}_2^{s,k-1} = \begin{bmatrix} -1 & 0 & 0 & 0 \\ 0 & -1 & 0 & 0 \\ -\dfrac{\sin\beta_{s,k}}{\mathrm{LD}_{s,k}} & \dfrac{\cos\beta_{s,k}}{\mathrm{LD}_{s,k}} & \dfrac{\sin\beta_{s,k}}{\mathrm{LD}_{s,k}} & -\dfrac{\cos\beta_{s,k}}{\mathrm{LD}_{s,k}} \end{bmatrix} \tag{2.2}$$

where $\mathrm{LD}_{s,k}$ is the distance between $\mathrm{LS}_{1,s,k}$ and $\mathrm{LS}_{2,s,k}$. At a current station, the total part/subassembly deviation is then the combination of current fixture location errors and reorientation errors. For a MAP, the resulting dimension variation can be estimated by recursively applying previous equations along all stages. The model structure for this variation propagation is taken from control theory as the well-known state-space model (Ogata and Yang, 2009). Then, the state-space model for a generic MAP with N-stations, as shown in Fig. 2.4, can be mathematically defined as

$$\boldsymbol{x}_k = \boldsymbol{A}_{k-1} \cdot \boldsymbol{x}_{k-1} + \boldsymbol{B}_k \cdot \boldsymbol{u}_k + \boldsymbol{w}_k \tag{2.3}$$

$$\boldsymbol{y}_k = \boldsymbol{C}_k \cdot \boldsymbol{x}_k + \boldsymbol{v}_k, \quad k = 1, \ldots, N \tag{2.4}$$

Eq. (2.3) is a state transition model, where $\boldsymbol{x}_k = \left[\boldsymbol{x}_{1,k}^T \cdots \boldsymbol{x}_{n,k}^T\right]^T \in \mathcal{R}^{3n\times 1}$ represents the deviations of quality features at stage k and n is the number of quality features considered in $\boldsymbol{x}_k$. $\boldsymbol{A}_k \in \mathcal{R}^{3n\times 3n}$ is the reorientation matrix that describes the error transferred from an upstream station through part/subassembly reorientation. $\boldsymbol{u}_k = [\delta LS_{1,s,k}(x) \cdots \delta LS_{2,s,k}(z)]^T \in \mathcal{R}^{2p\times 1}$ represents the deviations of the locators, e.g., $\mathrm{LS}_{1,s,k}$ and

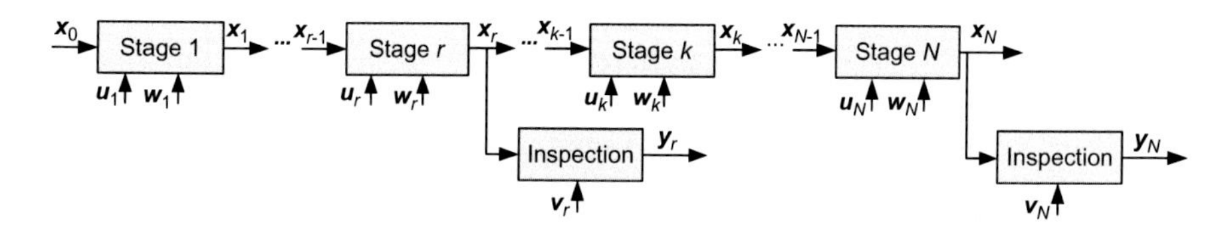

Fig. 2.4 Multistage manufacturing process.

$LS_{2,s,k}$, from their nominal positions, where p is the number of locators. $\boldsymbol{B}_k \in \mathcal{R}^{3n\times 2p}$ describes the impacts of locator deviations on $\boldsymbol{x}_k$. Eq. (2.4) is an observation model, where matrix $\boldsymbol{C}_k \in \mathcal{R}^{m\times 3n}$ represents how the deviations of quality features are transformed to the deviations of the measurements conducted at the features, which are defined as the key product characteristics (KPCs), and m is the number of measurement. These measurements are denoted by $\boldsymbol{y}_k = [y_{1,k} \cdots y_{m,k}]^T \in \mathcal{R}^{m\times 1}$. The unmodeled disturbance and the measurement errors are denoted as $\boldsymbol{w}_k = \left[\boldsymbol{w}_{1,k}^T \cdots \boldsymbol{w}_{n,k}^T\right]^T$ and $\boldsymbol{v}_k = [v_{1,k} \cdots v_{m,k}]^T$, respectively, and $\boldsymbol{w}_k \in \mathcal{R}^{3n\times 1}$ and $\boldsymbol{v}_k \in \mathcal{R}^{m\times 1}$.

The details of the complete mathematical derivation of Eqs. (2.3) and (2.4) are omitted for simplicity. Interested readers may refer to (Shi, 2006) for detailed model derivation procedures.

Please note that, without loss of generality, the work presented here will be focused on 2D MAP model. However, an extended version of the model in order to deal with 3D MAP, presented in Liu et al. (2010), could be also applied.

2.3 Feed-forward predictive control

Two basic approaches have been investigated in controlling variation magnitude in multistage manufacturing processes: feed-back and feed-forward control. Feed-back control makes use of information from downstream measurements to determine control actions at upstream stations for a next set of parts. For instance, after inspecting the quality of a sample of parts at the last station, a mean shift of a KPC may be detected. The corresponding control action for the next batch of parts to be manufactured is determined and performed at proper stations.

Alternatively, feed-forward control requires in-process measurements to predict and minimize final product deviations by acting at downstream stages. Unlike the feed-back control strategy, this strategy allows the dimensional adjustments for either a batch of parts or an individual part. Therefore feed-forward control is more preferable in variation reduction for multistage manufacturing processes. Both feed-back and feed-forward control strategies are shown in Fig. 2.5.

The first work in the field of feed-forward control for variation reduction in multistage manufacturing processes was conducted by Djurdjanovic and Zhu (2005). The feed-back and feed-forward control strategies for the placement of stations with dimensional adjustment capability were proposed. Innovatively addressing the dimension compensation problem, this work considers only the deterministic effects, neglecting the noise due to the linearization, unmodeled effects, process noise, and sensor imperfection. Furthermore, the concept of compensability was introduced to quantitatively evaluate the capability of variation compensation in a specific system. This concept is equivalent to the concept of controllability from control theory (Ogata and Yang, 2009) and enables the identification of the compensable variation sources. Izquierdo et al. (2007) extended the feed-forward control strategy to include parts/process requirements and specific engineering constraints on the magnitudes of control actions, such as physical limits and inaccuracy of tooling adjustments.

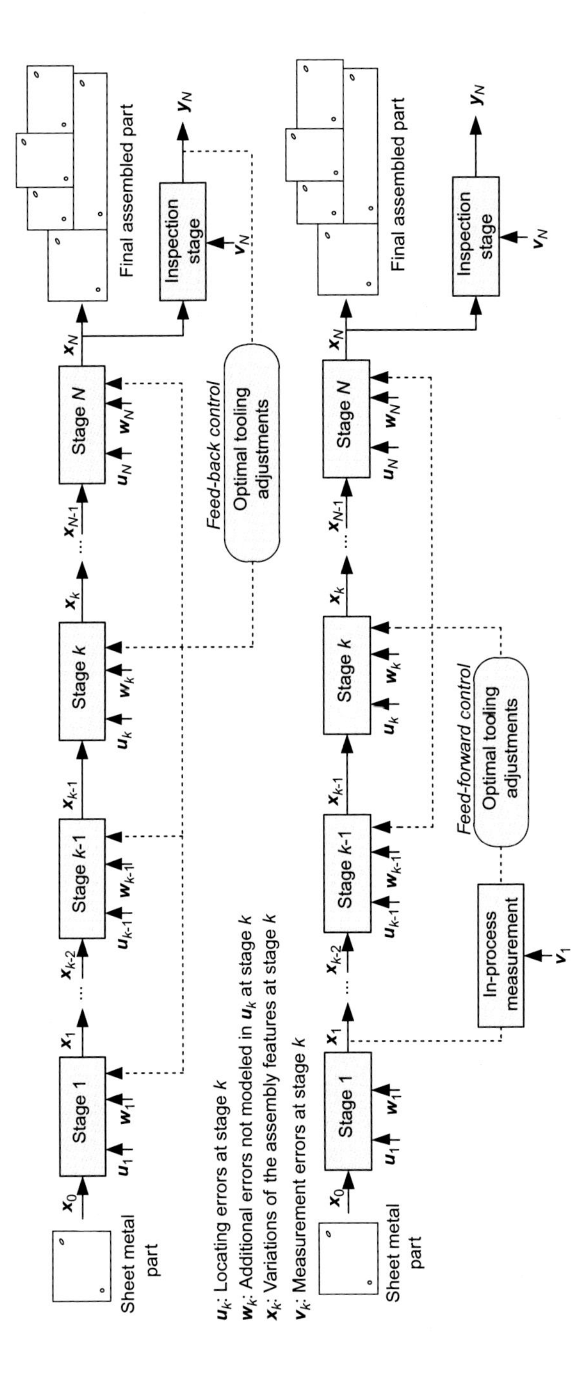

Fig. 2.5 Feed-back and feed-forward control strategies.

Previous works were focused on the study of feed-forward control with a full control of all tooling elements. This assumption may not be realistic, since tooling adjustments through flexible fixtures or CNC machine tools may only be deployed to selected stations in the system due to their high costs. Thus Djurdjanovic and Ni (2007) proposed a feed-forward control strategy with distributed actuation capabilities, taking into consideration the actuation accuracy and noise. However, they only select the best placement from the potential and distributed tooling adjustments, without considering the interaction of multiple tooling adjustments. Metaheuristic optimization approaches were used in Jiao and Djurdjanovic (2010), where the research work in Djurdjanovic and Ni (2007) was extended to deal with variation reduction considering multiple tooling adjustments. The placement of flexible fixtures at multiple stations was analyzed for adjusting tooling elements after multiple in-process measurements. The optimization of the flexible fixtures allocation in MAPs was conducted using a Tabu search algorithm. In Zhong et al. (2010) the impact of model uncertainty on the control performance in MAPs is investigated and a robust control action is proposed. However, the analysis of multiple controllers and multiple in-process measurement stations was not discussed. More recently, the feed-forward control was applied in multistage machining processes with sensor-based fixtures and CNC controllers (Abellán-Nebot et al., 2012).

In this chapter, a predictive control strategy with moving horizon is proposed. Under this control strategy, the full set of control actions to be applied in current and subsequent stages is obtained according to previous inspection measurements, but only the control action for current stage is applied. The process is repeated at subsequent stages, but incorporating new measurements which let update the full set of control actions.

The derivation of this control strategy at stage i is obtained by solving the following optimization problem: *given a set of measurements* y_{i-1} *, find the set of control actions* $\{\boldsymbol{u}_i, \boldsymbol{u}_{i+1}, \ldots, \boldsymbol{u}_N\}$ *that minimizes the expected value of* $\boldsymbol{y}_N^T \cdot \boldsymbol{y}_N$ *and apply control action* $\boldsymbol{u}_i$ *at the current stage i.*

Thus the following index should be optimized for each stage:

$$\min_{\boldsymbol{U}_i} J_i = \mathrm{E}\left\{(\boldsymbol{C}_N \cdot \boldsymbol{x}_N)^T \cdot (\boldsymbol{C}_N \cdot \boldsymbol{x}_N)\right\} \tag{2.5}$$

$$\text{s.t.} \begin{cases} \boldsymbol{C}_N \cdot \boldsymbol{x}_N = \boldsymbol{\Gamma}_i \cdot \boldsymbol{x}_{i-1} + \boldsymbol{\Psi}_i \cdot \boldsymbol{U}_i + \boldsymbol{\Omega}_i \cdot \boldsymbol{W}_i, \\ \boldsymbol{y}_{i-1} = \boldsymbol{C}_{i-1} \cdot \boldsymbol{x}_{i-1} + \boldsymbol{v}_{i-1} \end{cases}$$

where $\mathrm{E}\{\cdot\}$ refers to the expected value, and the earlier matrices are defined as follows

$$\boldsymbol{\Gamma}_i = \boldsymbol{C}_N \cdot \boldsymbol{A}_N \cdot \ldots \cdot \boldsymbol{A}_i$$

$$\boldsymbol{\Psi}_i = \boldsymbol{C}_N \cdot [\boldsymbol{A}_N \cdot \ldots \cdot \boldsymbol{A}_{i+1} \cdot \boldsymbol{B}_i,\ \boldsymbol{A}_N \cdot \ldots \cdot \boldsymbol{A}_{i+2} \cdot \boldsymbol{B}_{i+1}, \ldots, \boldsymbol{B}_N]$$

$$\boldsymbol{\Omega}_i = \boldsymbol{C}_N \cdot [\boldsymbol{A}_N \cdot \ldots \cdot \boldsymbol{A}_{i+1},\ \boldsymbol{A}_N \cdot \ldots \cdot \boldsymbol{A}_{i+2}, \ldots, \boldsymbol{I}_{n\times n}]$$

$$\mathbf{U}_i = \begin{bmatrix} \boldsymbol{u}_i \\ \vdots \\ \boldsymbol{u}_N \end{bmatrix}, \quad \mathbf{W}_i = \begin{bmatrix} \boldsymbol{w}_i \\ \vdots \\ \boldsymbol{w}_N \end{bmatrix}$$

These equations estimate the error propagation from stage $i-1$ to N as a function of the control actions and disturbances (i.e., modeling and measurement errors) that will apply throughout the MAP. In the Appendix it is described how to obtain the matrices $\boldsymbol{\Gamma}_i$, $\boldsymbol{\Psi}_i$, and $\boldsymbol{\Omega}_i$.

The solution to the previous control problem with moving horizon is

$$\boldsymbol{U}_i^* = -\left(\boldsymbol{\Psi}_i^T \cdot \boldsymbol{\Psi}_i\right)^{\dagger} \cdot \boldsymbol{\Psi}_i^T \cdot \boldsymbol{\Gamma}_i \cdot \boldsymbol{x}_{i-1} = \boldsymbol{K}_i^* \cdot \boldsymbol{x}_{i-1}$$

$$\boldsymbol{u}_i^* = \left[\boldsymbol{I}_{2p\times 2p} \quad \boldsymbol{0}_{2p\times 2p\cdot(N-i)}\right] \cdot \boldsymbol{U}_i^* = \boldsymbol{S}_i \cdot \boldsymbol{U}_i^* = \boldsymbol{S}_i \cdot \boldsymbol{K}_i^* \cdot \boldsymbol{x}_{i-1} = \boldsymbol{k}_i^* \cdot \boldsymbol{x}_{i-1} \tag{2.6}$$

where $(\cdot)^{\dagger}$denotes the Moore-Penrose inverse of a matrix, matrix $\boldsymbol{k}_i^*$ is used to represent compactly the solution of the control problem, and matrix $\boldsymbol{S}_i$ is used to select the control action to be applied (the first $2p$ values of vector $\boldsymbol{U}_i^*$). Please, refer to Appendix for derivation details.

2.4 Optimal inspection/actuator placement for feed-forward control

MAPs may be composed of a large number of stations, but not all stations have the ability of modifying its operations through a flexible tooling. Meanwhile, not all stations are equipped with in-process measurements to estimate the current state of the manufacturing process in terms of product dimensional quality. Therefore a process planning decision is required in order to define where to place the inspection and actuation stations (i.e., stations whose fixtures can be modified part by part through flexible tooling or similar). Fig. 2.6 shows the problem description.

In MAPs, the placement of a controller (e.g., programmable tooling) is more costly than the placement of an in-process measurement stations. For this reason, a two-step methodology is defined for the optimal inspection/actuator placement. In the first step, it is considered that an in-process measurement is available at any station of a MAP, and it is the placement of a controller, which should be minimized in order to ensure product specifications. Then, in the second step, based on the actuators available, and according to their placement, it minimizes the number of in-process measurements stations.

The mathematical formulation of this two-step methodology is described as follows.

2.4.1 Step 1: Optimal placement of actuators

In this step, it is assumed that there is an inspection at any station throughout the MAP. The optimal placement of actuators begins by evaluating the index from Eq. (2.5), considering that a controller is placed at all stations.

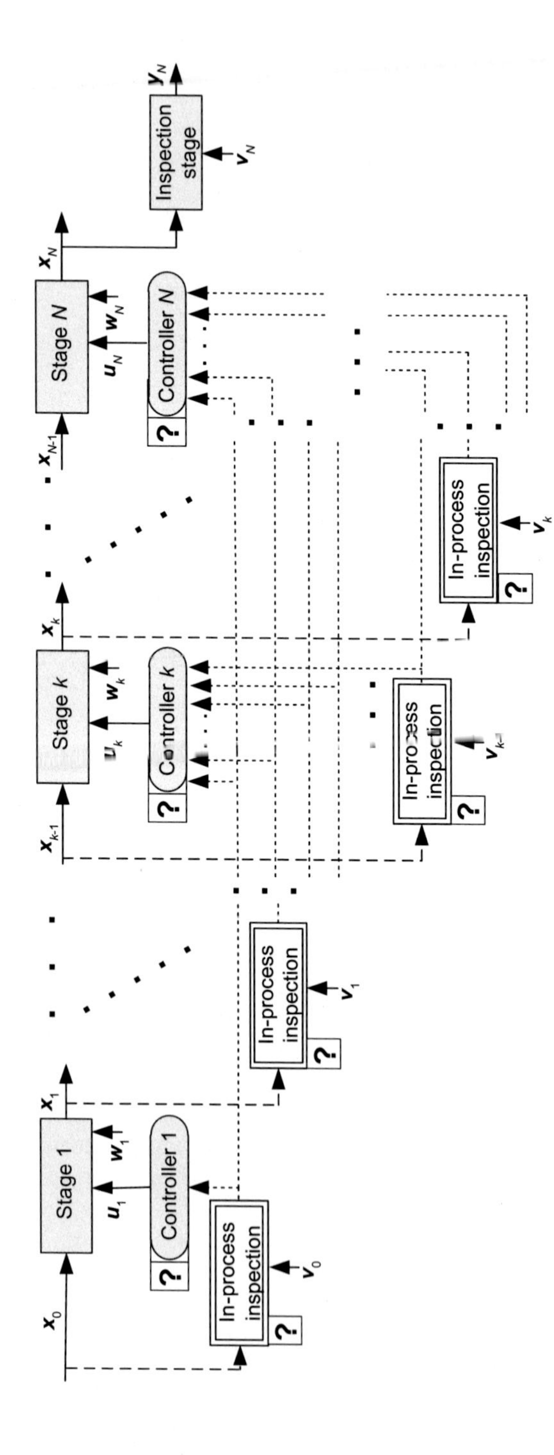

Fig. 2.6 Multistage manufacturing process with feed-forward control based on optimal inspection/actuator placement.

Then, in order to reduce the number of actuators, the algorithm identifies the actuator that, if removed, will increase the index in the smallest amount, while ensuring product specifications. This actuator will be set to zero in order to remove it from the MAP, and the procedure will be repeated until the removal of the less important actuator will result in products out of specifications. In that case, the optimal placement of actuators is given. Please note that, if an actuator is removed from a stage in the MAP, the inspection station placed before that stage will also be removed.

Therefore the index from Eq. (2.5) is rewritten as:

$$\min_{\boldsymbol{U}_i} J_i = \mathrm{E}\left\{ (\boldsymbol{C}_N \cdot \boldsymbol{x}_N)^T \cdot (\boldsymbol{C}_N \cdot \boldsymbol{x}_N) \right\} \tag{2.7}$$

$$\text{s.t.} \begin{cases} \boldsymbol{C}_N \cdot \boldsymbol{x}_N = \boldsymbol{\Gamma}_i \cdot \boldsymbol{x}_{i-1} + \boldsymbol{\Psi}_i \cdot \boldsymbol{U}_i + \boldsymbol{\Omega}_i \cdot \boldsymbol{W}_i \\ \boldsymbol{y}_{i-1} = \boldsymbol{C}_{i-1} \cdot \boldsymbol{x}_{i-1} + \boldsymbol{v}_{i-1} \\ u_j = 0, \text{ for some } j \in \{2 \cdot (i-1) \cdot p + 1, \ldots, 2 \cdot p \cdot N\} \end{cases}$$

where, at each stage, the control action is given by

$$\boldsymbol{U}_i = -\left(\boldsymbol{\Psi}_i^T \cdot \boldsymbol{\Psi}_i\right)^{-1} \cdot \left(\boldsymbol{I} - \boldsymbol{Q}_{m,i}^T \cdot \left(\boldsymbol{Q}_{m,i} \cdot \left(\boldsymbol{\Psi}_i^T \cdot \boldsymbol{\Psi}_i\right)^{-1} \cdot \boldsymbol{Q}_{m,i}^T \right)^{-1} \cdot \left(\boldsymbol{\Psi}_i^T \cdot \boldsymbol{\Psi}_i\right)^{-1} \right) \cdot$$
$$\boldsymbol{\Psi}_i^T \cdot \boldsymbol{\Gamma}_i \cdot \boldsymbol{x}_{i-1} = \boldsymbol{K}_i \cdot \boldsymbol{x}_{i-1}$$

$$\boldsymbol{u}_i = \left[\boldsymbol{I}_{2p \times 2p} \quad \boldsymbol{0}_{2p \times 2p \cdot (N-i)} \right] \cdot \boldsymbol{U}_i = \boldsymbol{S}_i \cdot \boldsymbol{U}_i = \boldsymbol{S}_i \cdot \boldsymbol{K}_i \cdot \boldsymbol{x}_{i-1} = \boldsymbol{k}_i \cdot \boldsymbol{x}_{i\;1} \tag{2.8}$$

In previous equations, matrix $\boldsymbol{Q}_{m,i}$ is a matrix of size $n_m \times 2(N-i+1) \cdot p$ which is used to indicate which control actions are set to zero, m is a set of actuator indices set to zero, and the number of elements is n_m. Matrix $\boldsymbol{Q}_{m,1}$ is formed by as many rows as number of actuators to be set to zero, and $2 \cdot p \cdot N$ columns. For instance, if actuators 2, 5, and 6 ($m = \{2, 5, 6\}$) are set to zero in a process with $N = 2$ stations and $p = 2$ possible actuator locations per station, matrix $\boldsymbol{Q}_{\{2,5,6\},1}$ will be

$$\boldsymbol{Q}_{\{2,5,6\},1} = \begin{bmatrix} 0 & 1 & 0 & 0 & 0 & 0 & 0 & 0 \\ 0 & 0 & 0 & 0 & 1 & 0 & 0 & 0 \\ 0 & 0 & 0 & 0 & 0 & 1 & 0 & 0 \end{bmatrix}$$

which leads to $\boldsymbol{Q}_{\{2,5,6\},1} \cdot \boldsymbol{U}_1 = \begin{bmatrix} u_2 \\ u_5 \\ u_6 \end{bmatrix}$, and the corresponding matrix for the next station is

$$\boldsymbol{Q}_{\{2,5,6\},2} = \begin{bmatrix} 0 & 0 & 0 & 0 \\ 1 & 0 & 0 & 0 \\ 0 & 1 & 0 & 0 \end{bmatrix}$$

and thus $\boldsymbol{Q}_{\{2,5,6\},2} \cdot \boldsymbol{U}_2 = \begin{bmatrix} u_5 \\ u_6 \end{bmatrix}$. Any matrix $\boldsymbol{Q}_{m,i}$ is obtained by extracting the last $2(N-i+1) \cdot p$ columns of $\boldsymbol{Q}_{m,1}$:

$$\boldsymbol{Q}_{m,i} = \boldsymbol{Q}_{m,1} \cdot \begin{bmatrix} \boldsymbol{0}_{2(i-1)p \times 2(N-i)p} \\ \boldsymbol{I}_{2(N-i)p \times 2(N-i)p} \end{bmatrix}$$

The implementation of this control action requires the knowledge of the dimensions at the previous station $\boldsymbol{x}_{i-1}$. This value is estimated from the measurements at the previous station as $\hat{\boldsymbol{x}}_{i-1} = \boldsymbol{C}_{i-1}^{\dagger} \cdot \boldsymbol{y}_{i-1}$. Therefore the control action can be expressed as a function of the real state $\boldsymbol{x}_{i-1}$ and the noise measurement $\boldsymbol{v}_{i-1}$ as

$$\boldsymbol{u}_i = \boldsymbol{k}_i \cdot \left(\boldsymbol{x}_{i-1} + \boldsymbol{C}_{i-1}^{\dagger} \cdot \boldsymbol{v}_{i-1} \right) \tag{2.9}$$

Applying the resulting control action at each station, the final cost index for all KPCs at the end of the MAP is defined as

$$J = trace\left(\boldsymbol{\Sigma}_{X_0} \cdot \boldsymbol{M}_x^T \cdot \boldsymbol{M}_x\right) + trace\left(\boldsymbol{\Sigma}_{W_1} \cdot \boldsymbol{M}_w^T \cdot \boldsymbol{M}_w\right) + trace\left(\boldsymbol{\Sigma}_{V_1} \cdot \boldsymbol{M}_v^T \cdot \boldsymbol{M}_v\right) \tag{2.10}$$

where $\boldsymbol{\Sigma}_{X_0}$, $\boldsymbol{\Sigma}_{W_1}$, and $\boldsymbol{\Sigma}_{V_1}$ are the covariance matrix for $\boldsymbol{x}_0$, $\boldsymbol{W}_1$, and $\boldsymbol{V}_1$, respectively, being

$$\mathbf{W}_1 = \begin{bmatrix} \boldsymbol{w}_1 \\ \vdots \\ \boldsymbol{w}_N \end{bmatrix}, \quad \mathbf{V}_1 = \begin{bmatrix} \boldsymbol{v}_1 \\ \vdots \\ \boldsymbol{v}_N \end{bmatrix}$$

and matrices $\boldsymbol{M}_x$, $\boldsymbol{M}_w$, and $\boldsymbol{M}_v$ are given by

$$\boldsymbol{M}_x = \boldsymbol{C}_N \cdot \boldsymbol{M}_{1:N}$$

$$\boldsymbol{M}_w = \boldsymbol{C}_N \cdot \left[\boldsymbol{M}_{2:N} \quad \cdots \quad \boldsymbol{M}_{N:N} \quad \boldsymbol{I}_{n \times n} \right]$$

$$\boldsymbol{M}_v = \boldsymbol{C}_N \cdot \left[\boldsymbol{M}_{2:N} \cdot \boldsymbol{B}_1 \cdot \boldsymbol{k}_1 \cdot \boldsymbol{C}_0^{\dagger} \quad \cdots \quad \boldsymbol{M}_{N:N} \cdot \boldsymbol{B}_{N-1} \cdot \boldsymbol{k}_{N-1} \cdot \boldsymbol{C}_{N-1}^{\dagger} \quad \boldsymbol{B}_N \cdot \boldsymbol{k}_N \cdot \boldsymbol{C}_N^{\dagger} \right]$$

with

$$\boldsymbol{M}_{i:N} = (\boldsymbol{A}_N + \boldsymbol{B}_N \cdot \boldsymbol{k}_N) \cdots (\boldsymbol{A}_{i+1} + \boldsymbol{B}_{i+1} \cdot \boldsymbol{k}_{i+1}) \cdot (\boldsymbol{A}_i + \boldsymbol{B}_i \cdot \boldsymbol{k}_i)$$

Please, refer to Appendix for derivation details.

2.4.2 *Step 2: Optimal placement of inspection stations*

In this step, given a specific number of controllers placed in a MAP, it is desired to minimize the current number of inspection stations while ensuring product specifications. For this purpose, the cost index is evaluated under the assumption that an inspection station is removed. The inspection station that less increases the cost index is chosen to be removed only if product quality can still be kept within specifications. The procedure will be repeated until the removal of the less important inspection station will result in products out of specifications. In that case, the optimal placement of inspection stations is given.

To evaluate the increase of the cost index when an inspection station is removed, the cost index is evaluated under two different situations: the one obtained using the control action $\boldsymbol{u}_i$ computed at the actual station with measurements $\boldsymbol{y}_{i-1}$, and the one obtained using the control action computed from the previous station with measurements $\boldsymbol{y}_{i-2}$, which means that the inspection station $\boldsymbol{y}_{i-1}$ has been removed. The control action predicted from a previous station is given by

$$\boldsymbol{u}_{i|i-1} = \left[\boldsymbol{0}_{2p\times 2p}\ \boldsymbol{I}_{2p\times 2p}\ \ \boldsymbol{0}_{2p\times 2p\cdot(N-i)}\right]\cdot \boldsymbol{U}_{i-1} = \boldsymbol{S}_{i|i-1}\cdot \boldsymbol{U}_{i-1} \tag{2.11}$$

For a general formulation when c consecutive measurement stations are removed, the control action applied is the one predicted from c upward stations and it is defined as

$$\boldsymbol{u}_{i|i-c} = \left[\boldsymbol{0}_{2p\times c\cdot 2p}\ \boldsymbol{I}_{2p\times 2p}\ \ \boldsymbol{0}_{2p\times 2p\cdot(N-i)}\right]\cdot \boldsymbol{U}_{i-c} = \boldsymbol{S}_{i|i-c}\cdot \boldsymbol{U}_{i-c} \tag{2.12}$$

2.5 Case study

A case study is presented to validate the proposed methodology. Fig. 2.7 shows a four-stage assembly process that assembles four parts at three consecutive stages and measures the final product KPCs at the fourth stage. The assembly operations, locating schemes, and measuring strategy are summarized in Tables 2.1 and 2.2. In the first stage, part 1 and part 2 are located at points $\{P_1, P_2\}$ and $\{P_3, P_4\}$, respectively. These two parts are assembled to form subassembly 1&2. The final assembly, 1&2&3&4, is located at the 4th stage to collect the measurements on X-direction and Z-direction deviations of the eight points, M1–M8. Thus the number of KPCs contained in y_4 is 16, i.e., $m=16$. Note that according to Table 2.1, there are four locating points at each station, and each locator has two position coordinates X and Z that could be actuated by a programmable tooling. Thus 24 potential actuators can be placed in this MAP.

From shop-floor data, the dimensional variations of sheet metal panels due to machining deviations in manufacturing the four-way hole and two-way slot can be

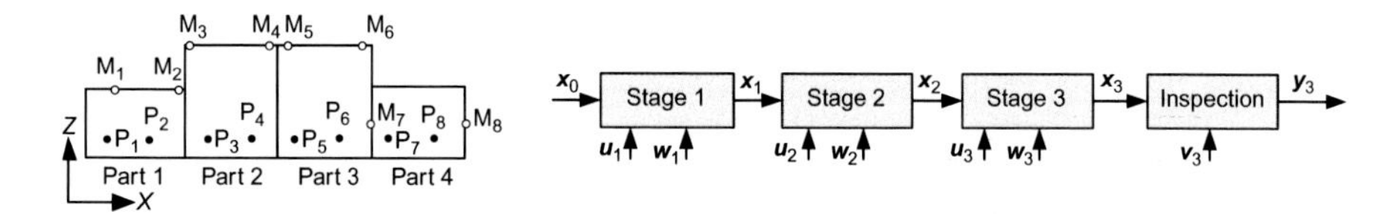

Fig. 2.7 Assembly process of the case study.

Table 2.1 Assembly operations and locating scheme

Stage index	Operations	Locating scheme 1		Locating scheme 2		Output subassembly
		Part/ subassembly	Locating points	Part/ subassembly	Locating points	
1	Assembly	1	$\{P_1,P_2\}$	2	$\{P_3,P_4\}$	1 & 2
2	Assembly	1 & 2	$\{P_1,P_4\}$	3	$\{P_5,P_6\}$	1 & 2 & 3
3	Assembly	1 & 2 & 3	$\{P_1,P_6\}$	4	$\{P_7,P_8\}$	1 & 2 & 3 & 4
4	Measuring	1 & 2 & 3 & 4	$\{P_1,P_8\}$	–	–	–

Table 2.2 Product and process design information (unit: mm)

Fixture locating pin positions			Measurement point positions		
	X	*Z*		*X*	*Z*
P_1	100	100	M_1	200	400
P_2	150	100	M_2	700	400
P_3	800	100	M_3	700	600
P_4	850	100	M_4	1500	600
P_5	1500	100	M_5	1550	600
P_6	1550	100	M_6	2100	600
P_7	2300	100	M_7	2200	200
P_8	2350	100	M_8	2700	200

estimated to place each part onto the fixture. For this case study, it is assumed that the sheet metal deviations from previous processes, defined by $\boldsymbol{x}_0$, follow a normal distribution with 0 mean and 0.05 mm standard deviation at coordinates *X* and *Z*, and a normal distribution with 0 mean and 0.002 rad standard deviation at β coordinate.

The quality department sets the maximum permissible deviation of any point of the sheet panel structure to ± 3 mm. In order to ensure product quality specifications and to optimally place the actuation and inspection units in the multistage manufacturing process, the methodology shown earlier is applied. Please note that, for the sake of simplicity, the measurement noise for each stage inspection is assumed to be the same, and it is defined by a normal distribution with 0 mean and 0.01 mm standard deviation.

The results of the proposed methodology are shown in Fig. 2.8. For illustrative purposes, the figure shows the maximum deviation in $\pm 3\sigma$ of any of the KPCs and three different MAP configurations: a MAP without any actuator or in-process inspection, a MAP where an actuator and an in-process inspection is placed at any station, which means a total control of the process, and a MAP where the placement of actuators and in-process inspection stations are minimized while ensuring product specifications.

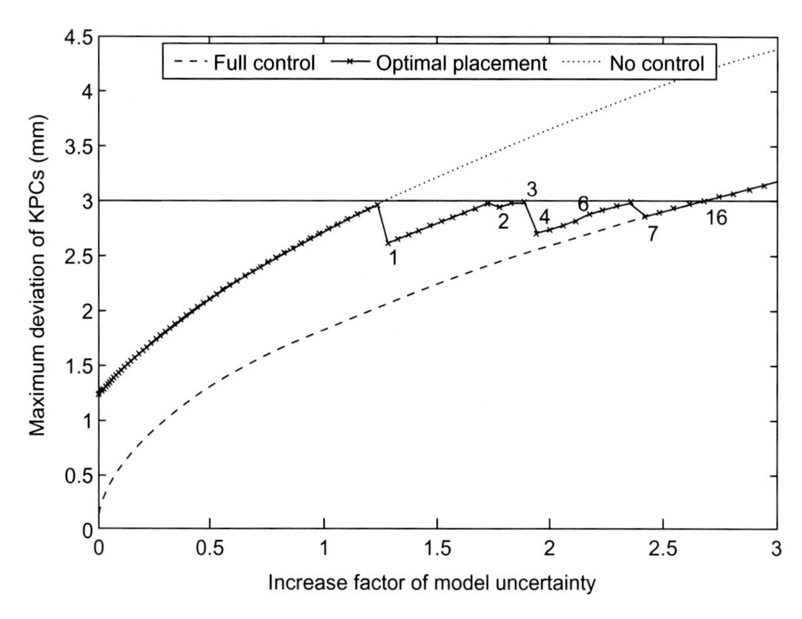

Fig. 2.8 Maximum deviations of KPCs according to different model uncertainty scenarios for three MAP configurations: no actuation/inspection units, full implementation of actuation/inspection stations, and optimal placement of actuation/inspection stations.

An increase factor of model uncertainty is used to illustrate the process performance at each configuration. Note that this factor multiplies the variance of model uncertainty which is initially set to $(0.05/3)^2$ mm^2 and $(0.002/3)^2$ rad^2.

It can be seen that, when model uncertainty is very low, all three configurations ensure product specifications. The MAP without any actuator stations manufacture products within specifications because the initial deviations of the raw material are low and the deviations added at each stations due to model uncertainty are also low. However, when model uncertainty increases, part deviations are increased and a feed-forward control approach is required. For an increase factor of 1.3, only one actuator is needed to ensure product specifications, which is placed at the third station (actuator number 20). Higher model uncertainty requires a higher number of actuators and inspection stations to ensure product specifications. As shown in Fig. 2.8, 6 and 7 actuators are required for an increase factor of 2.1 and 2.4, respectively. When model uncertainty is increased by a factor of 2.7, it is not possible to ensure product specifications even though all stations would present an actuator and inspection unit.

The results of the methodology presented earlier to place optimally the actuator stations are shown in Fig. 2.9. The graph shows, for an increase factor of model uncertainty of 1.8, how the algorithm removes the less important station for controlling purposes, repeating the process until product quality may be out of specifications. For this case, from 24 possible actuators (i.e., 8 possible actuators for each station), the optimal number of actuators is 2, which are defined by the actuators {16, 24}.

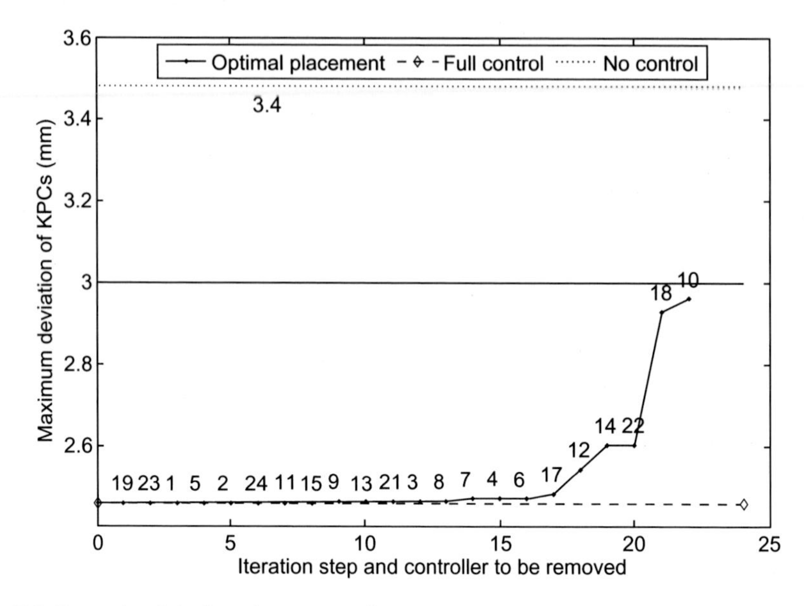

Fig. 2.9 Example of the iterative process for optimal actuator placement. The less important actuator is removed at each iteration (the actuator number is shown in the graph). Note that the maximum deviations of KPCs are always kept lower than ± 3 mm.

2.6 Conclusions

Nowadays, multistage manufacturing systems are expected to be capable of adapting themselves from changes or deviations along the production line in order to keep product quality within certain specification levels. For this purpose, process knowledge to estimate the performance of current manufacturing operations and advanced control schemes to properly actuate in subsequent stages have to be applied by engineers.

In this chapter, the state-space model for modeling multistage manufacturing processes was briefly described and its application for feed-forward control to compensate manufacturing errors at downstream stations was discussed. Given a product quality specification, a new methodology is derived to analyze where to place in-process measurement stations to estimate current process state and where to place stations to adapt their parameters according to previous estimations. The optimization is conducted under a step-based procedure and taking into account model uncertainty in order to ensure a robust dimensional compensation.

The proposed methodology was applied in a MAP in order to validate the effectiveness of the optimization process under different values of model uncertainty. The simulations showed the impact of model uncertainty in the controlling scheme and the necessity of different MAP configurations with different actuation/inspection stations to ensure product quality specifications.

Appendix

Derivation of matrices Γ_i, Ψ_i, and Ω_i

At the last station, the inspected KPCs $\boldsymbol{y}_N$ are related to the deviation of the sheet metal panels as

$$\boldsymbol{y}_N = \boldsymbol{C}_N \cdot \boldsymbol{x}_N$$

It should be noted that $\boldsymbol{x}_N$ is the result of the sheet metal assembly process at station N given the subassembly deviations $\boldsymbol{x}_{N-1}$, then

$$\boldsymbol{x}_N = \boldsymbol{A}_N \cdot \boldsymbol{x}_{N-1} + \boldsymbol{B}_N \cdot \boldsymbol{u}_N + \boldsymbol{w}_N$$

Additionally, $\boldsymbol{x}_{N-1}$ is the result of moving $\boldsymbol{x}_{N-2}$ through station $N-1$, leading to

$$\boldsymbol{x}_{N-1} = \boldsymbol{A}_{N-1} \cdot \boldsymbol{x}_{N-2} + \boldsymbol{B}_{N-1} \cdot \boldsymbol{u}_{N-1} + \boldsymbol{w}_{N-1}$$

Therefore $\boldsymbol{x}_N$ can be written as a function of $\boldsymbol{x}_{N-2}$ as

$$\boldsymbol{x}_N = \boldsymbol{A}_N \cdot (\boldsymbol{A}_{N-1} \cdot \boldsymbol{x}_{N-2} + \boldsymbol{B}_{N-1} \cdot \boldsymbol{u}_{N-1} + \boldsymbol{w}_{N-1}) + \boldsymbol{B}_N \cdot \boldsymbol{u}_N + \boldsymbol{w}_N$$

$$\boldsymbol{x}_N = \boldsymbol{A}_N \cdot \boldsymbol{A}_{N-1} \cdot \boldsymbol{x}_{N-2} + \begin{bmatrix} \boldsymbol{A}_N \cdot \boldsymbol{B}_{N-1} & \boldsymbol{B}_N \end{bmatrix} \begin{bmatrix} \boldsymbol{u}_{N-1} \\ \boldsymbol{u}_N \end{bmatrix} + \begin{bmatrix} \boldsymbol{A}_N & \boldsymbol{I}_{n \times n} \end{bmatrix} \begin{bmatrix} \boldsymbol{w}_{N-1} \\ \boldsymbol{w}_N \end{bmatrix}$$

where $\boldsymbol{I}_{n \times n}$ is the identity matrix of size n.

Applying this idea recursively, $\boldsymbol{x}_N$ is defined as

$$\begin{aligned} \boldsymbol{x}_N = \boldsymbol{A}_N \cdots \boldsymbol{A}_i \cdot \boldsymbol{x}_{i-1} + \begin{bmatrix} \boldsymbol{A}_N \cdots \boldsymbol{A}_{i+1} \cdot \boldsymbol{B}_i & \cdots & \boldsymbol{A}_N \cdot \boldsymbol{B}_{N-1} & \boldsymbol{B}_N \end{bmatrix} \cdot \boldsymbol{U}_i + \\ + \begin{bmatrix} \boldsymbol{A}_N \cdots \boldsymbol{A}_{i+1} & \cdots & \boldsymbol{A}_N & \boldsymbol{I}_{n \times n} \end{bmatrix} \cdot \boldsymbol{W}_i \end{aligned}$$

Finally, the KPCs at station N are obtained premultiplying the previous equation by $\boldsymbol{C}_N$, leading to

$$\boldsymbol{C}_N \cdot \boldsymbol{x}_N = \boldsymbol{\Gamma}_i \cdot \boldsymbol{x}_{i-1} + \boldsymbol{\Psi}_i \cdot \boldsymbol{U}_i + \boldsymbol{\Omega}_i \cdot \boldsymbol{W}_i$$

Derivation of the unconstrained optimal control action

The optimal control actions should be those that minimize the variability of the KPCs at the last station. The KPCs variability estimated at station i is defined as the cost index by the expression

$$\begin{aligned} J_i &= \mathrm{E}\left\{ (\boldsymbol{C}_N \cdot \boldsymbol{x}_N)^T \cdot (\boldsymbol{C}_N \cdot \boldsymbol{x}_N) \right\} \\ &= \mathrm{E}\left\{ (\boldsymbol{\Gamma}_i \cdot \boldsymbol{x}_{i-1} + \boldsymbol{\Psi}_i \cdot \boldsymbol{U}_i + \boldsymbol{\Omega}_i \cdot \boldsymbol{W}_i)^T (\boldsymbol{\Gamma}_i \cdot \boldsymbol{x}_{i-1} + \boldsymbol{\Psi}_i \cdot \boldsymbol{U}_i + \boldsymbol{\Omega}_i \cdot \boldsymbol{W}_i) \right\} \end{aligned}$$

Under the assumption of uncorrelated values for $\boldsymbol{x}_{i-1}$, $\boldsymbol{U}_i$, $\boldsymbol{W}_i$, and the zero mean assumption for $\boldsymbol{W}_i$, the previous expression can be rewritten as

$$J_i = (\boldsymbol{\Gamma}_i \cdot \boldsymbol{x}_{i-1} + \boldsymbol{\Psi}_i \cdot \boldsymbol{U}_i)^T (\boldsymbol{\Gamma}_i \cdot \boldsymbol{x}_{i-1} + \boldsymbol{\Psi}_i \cdot \boldsymbol{U}_i) + trace\left(\mathrm{E}\left\{ \boldsymbol{W}_i \cdot \boldsymbol{W}_i^T \right\} \cdot \boldsymbol{\Omega}_i^T \cdot \boldsymbol{\Omega}_i\right)$$

The optimal control action will set the derivative $\frac{\partial J_i}{\partial \boldsymbol{U}_i}$ to $\mathbf{0}$. The derivation of the cost index is defined as

$$\frac{\partial J_i}{\partial \boldsymbol{U}_i} = 2\boldsymbol{\Psi}_i^T \cdot (\boldsymbol{\Psi}_i \cdot \boldsymbol{U}_i + \boldsymbol{\Gamma}_i \cdot \boldsymbol{x}_{i-1}) = 0$$

The optimal control action will depend on the number of actuators and KPCs. If the number of actuators is higher than the number of KPCs, previous equation is an undetermined system and the minimum norm solution is given by

$$\boldsymbol{U}_i^* = -\boldsymbol{\Psi}_i^T \cdot \left(\boldsymbol{\Psi}_i \cdot \boldsymbol{\Psi}_i^T\right)^{-1} \cdot \boldsymbol{\Gamma}_i \cdot \boldsymbol{x}_{i-1}$$

If the number of actuators is lower than the number of KPCs, the optimal control action is

$$\boldsymbol{U}_i^* = -\left(\boldsymbol{\Psi}_i^T \cdot \boldsymbol{\Psi}_i\right)^{-1} \cdot \boldsymbol{\Psi}_i^T \cdot \boldsymbol{\Gamma}_i \cdot \boldsymbol{x}_{i-1}$$

A general expression can be applied for both cases if the pseudo-inverse operator is included thanks to its properties. Then, the optimal control action is defined as

$$\boldsymbol{U}_i^* = -\left(\boldsymbol{\Psi}_i^T \cdot \boldsymbol{\Psi}_i\right)^{\dagger} \cdot \boldsymbol{\Psi}_i^T \cdot \boldsymbol{\Gamma}_i \cdot \boldsymbol{x}_{i-1}$$

Derivation of the constrained optimal control action

In order to obtain the optimal control action forcing some null control actions to explore its effect on the final cost index, the problem is reformulated as follows

$$\min_{\boldsymbol{U}_i} J_i = \mathrm{E}\left\{ (\boldsymbol{C}_N \cdot \boldsymbol{x}_N)^T \cdot (\boldsymbol{C}_N \cdot \boldsymbol{x}_N) \right\}$$

$$\text{s.t.} \begin{cases} \boldsymbol{C}_N \cdot \boldsymbol{x}_N = \boldsymbol{\Gamma}_i \cdot \boldsymbol{x}_{i-1} + \boldsymbol{\Psi}_i \cdot \boldsymbol{U}_i + \boldsymbol{\Omega}_i \cdot \boldsymbol{W}_i \\ \boldsymbol{y}_{i-1} = \boldsymbol{C}_{i-1} \cdot \boldsymbol{x}_{i-1} + \boldsymbol{v}_{i-1} \\ \boldsymbol{Q}_{m,i} \cdot \boldsymbol{U}_i = 0 \end{cases}$$

To solve the previous optimization problem the method of Lagrange multipliers is applied. The Lagrangian for this problem is

$$\mathcal{L}(\boldsymbol{U}_i, \boldsymbol{\lambda}_i) = J_i + \boldsymbol{\lambda}_i^T \cdot \boldsymbol{Q}_{m,i} \cdot \boldsymbol{U}_i$$

and the equations to be solved are

$$\frac{\partial \mathcal{L}(\boldsymbol{U}_i, \boldsymbol{\lambda}_i)}{\partial \boldsymbol{U}_i} = 0, \quad \frac{\partial \mathcal{L}(\boldsymbol{U}_i, \boldsymbol{\lambda}_i)}{\partial \boldsymbol{\lambda}_i} = 0$$

Taking derivatives and solving the system of equations leads to

$$\boldsymbol{Q}_{m,i} \cdot \boldsymbol{U}_i = 0, \quad 2 \cdot \boldsymbol{\Psi}_i^T \cdot (\boldsymbol{\Psi}_i \cdot \boldsymbol{U}_i + \boldsymbol{\Gamma}_i \cdot \boldsymbol{x}_{i-1}) + \boldsymbol{Q}_{m,i}^T \cdot \boldsymbol{\lambda}_i = 0$$

$$\boldsymbol{\lambda}_i = -2 \cdot \boldsymbol{Q}_{m,i} \cdot \boldsymbol{\Psi}_i^T \cdot (\boldsymbol{\Gamma}_i \cdot \boldsymbol{x}_{i-1} + \boldsymbol{\Psi}_i \cdot \boldsymbol{U}_i)$$

$$\boldsymbol{U}_i = -\left(\boldsymbol{\Psi}_i^T \cdot \boldsymbol{\Psi}_i\right)^{\dagger} \cdot \left(\boldsymbol{\Psi}_i^T \cdot \boldsymbol{\Gamma}_i \cdot \boldsymbol{x}_{i-1} + \frac{1}{2} \cdot \boldsymbol{Q}_{m,i}^T \cdot \boldsymbol{\lambda}_i\right)$$

where $\boldsymbol{Q}_{m,i} \cdot \boldsymbol{Q}_{m,i}^T = \boldsymbol{I}_{m \times m}$. The resulting expressions for both $\boldsymbol{\lambda}_i$ and $\boldsymbol{U}_i$ are

$$\boldsymbol{\lambda}_i = 2 \cdot \left(\boldsymbol{Q}_{m,i} \cdot \left(\boldsymbol{\Psi}_i^T \cdot \boldsymbol{\Psi}_i\right)^{\dagger} \cdot \boldsymbol{Q}_{m,i}^T\right)^{\dagger} \cdot \boldsymbol{Q}_{m,i} \cdot \left(\boldsymbol{\Psi}_i^T \cdot \boldsymbol{\Psi}_i\right)^{\dagger} \cdot \boldsymbol{\Psi}_i^T \cdot \boldsymbol{\Gamma}_i \cdot \boldsymbol{x}_{i-1}$$

$$\begin{aligned}\boldsymbol{U}_i = &-\left(\boldsymbol{\Psi}_i^T \cdot \boldsymbol{\Psi}_i\right)^{\dagger} \cdot \left(\boldsymbol{I}_{(N-i+1)\cdot 2p \times (N-i+1)\cdot 2p} - \boldsymbol{Q}_{m,i}^T \cdot \left(\boldsymbol{Q}_{m,i} \cdot \left(\boldsymbol{\Psi}_i^T \cdot \boldsymbol{\Psi}_i\right)^{\dagger} \cdot \boldsymbol{Q}_{m,i}^T\right)^{\dagger} \cdot \right. \\ &\left. \boldsymbol{Q}_{m,i} \cdot \left(\boldsymbol{\Psi}_i^T \cdot \boldsymbol{\Psi}_i\right)^{\dagger}\right) \cdot \boldsymbol{\Psi}_i^T \cdot \boldsymbol{\Gamma}_i \cdot \boldsymbol{x}_{i-1} = \boldsymbol{k}_i \cdot \boldsymbol{x}_{i-1}\end{aligned}$$

Derivation of final state after control implementation

In order to obtain cost index J, the state $\boldsymbol{x}_N$ should be derived as a function of the applied control actions. For a generic station i, $\boldsymbol{x}_i$ is defined as

$$\begin{aligned}\boldsymbol{x}_i &= \boldsymbol{A}_i \cdot \boldsymbol{x}_{i-1} + \boldsymbol{B}_i \cdot \boldsymbol{k}_i \cdot \boldsymbol{x}_{i-1} + \boldsymbol{B}_i \cdot \boldsymbol{k}_i \cdot \boldsymbol{C}_{i-1}^{\dagger} \cdot \boldsymbol{v}_{i-1} + \boldsymbol{w}_i \\ &= (\boldsymbol{A}_i + \boldsymbol{B}_i \cdot \boldsymbol{k}_i) \cdot \boldsymbol{x}_{i-1} + \boldsymbol{B}_i \cdot \boldsymbol{k}_i \cdot \boldsymbol{C}_{i-1}^{\dagger} \cdot \boldsymbol{v}_{i-1} + \boldsymbol{w}_i\end{aligned}$$

Applying this expression recursively from $i = N$ to $i = 1$, $\boldsymbol{x}_N$ is derived as

$$\boldsymbol{x}_N = \prod_{i=1}^{N} (\boldsymbol{A}_i + \boldsymbol{B}_i \cdot \boldsymbol{k}_i) \cdot \boldsymbol{x}_0 + \sum_{i=1}^{N} \left(\prod_{j=i+1}^{N} (\boldsymbol{A}_j + \boldsymbol{B}_j \cdot \boldsymbol{k}_j)\right) \left(\boldsymbol{w}_i + \boldsymbol{B}_i \cdot \boldsymbol{k}_i \cdot \boldsymbol{C}_{i-1}^{\dagger} \cdot \boldsymbol{v}_{i-1}\right)$$

where the new matrices in the product must be left multiplied. That allows us to express $\boldsymbol{x}_N$ as a function of the initial raw material $\boldsymbol{x}_0$ and the disturbances $\boldsymbol{w}_i$ and noises $\boldsymbol{v}_{i-1}$ $(i = 1, \ldots, N)$.

References

Abellán-Nebot, J.V., Liu, J., Romero Subirón, F., 2012. Quality prediction and compensation in multi-station machining processes using sensor-based fixtures. Robot. Comput. Integr. Manuf. 28 (2), 208–219. http://dx.doi.org/10.1016/j.rcim.2011.09.001.

Djurdjanovic, D., Ni, J., 2007. Online stochastic control of dimensional quality in multistation manufacturing systems. Proc. Inst. Mech. Eng. B J. Eng. Manuf. 221 (5), 865–880. http://dx.doi.org/10.1243/09544054JEM458.

Djurdjanovic, D., Zhu, J., 2005. Stream of variation based error compensation strategy in multi-stage manufacturing processes. In: Manufacturing Engineering and Materials Handling, Parts A and B, vol. 2005, Proceedings, ASME, pp. 1223–1230. http://dx.doi.org/10.1115/IMECE2005-81550.

EFFRA, 2013. Factories of the future—multi-annual roadmap for the contractual PPP under Horizon 2020. European Factories of the Future Research Association (EFFRA). https:/doi.org/10.2777/29815.

IMS2020, 2010. Roadmap on sustainable manufacturing. Energy efficient manufacturing and key technologies.

Izquierdo, L.E., Shi, J., Hu, S.J., Wampler, C.W., 2007. Feedforward control of multistage assembly processes using programmable tooling. Trans. NAMRI/SME 35, 295–302.

Jiao, Y., Djurdjanovic, D., 2010. Joint allocation of measurement points and controllable tooling machines in multistage manufacturing processes. IIE Trans. 42 (10), 703–720. http://dx.doi.org/10.1080/07408170903544330.

Liu, J., 2010. Variation reduction for multistage manufacturing processes: a comparison survey of statistical-process-control vs stream-of-variation methodologies. Qual. Reliab. Eng. Int. 26, 645–661. http://dx.doi.org/10.1002/qre.1148.

Liu, J., Jin, J., Shi, J., 2010. State space modelling for 3-D variation propagation in rigid-body multistage assembly processes. IEEE Trans. Autom. Sci. Eng. 7 (2), 274–290. http://dx.doi.org/10.1109/TASE.2009.2012435.

Montgomery, D.C., 2007. Introduction to Statistical Quality Control. John Wiley & Sons, Hoboken, NJ.

Niaki, S.T.A., Davoodi, M., 2009. Designing a multivariate-multistage quality control system using artificial neural networks. Int. J. Prod. Res. 47 (1), 251–271. http://dx.doi.org/10.1080/00207540701504348.

Ogata, K., Yang, Y., 2009. Modern Control Engineering, fifth ed. Pearson, Upper Saddle River, NJ.

Paik, Y.J., 2008. Adjusting manufacturing process control parameter using updated process threshold derived from uncontrollable error. misc, Google Patents. Retrieved from https://www.google.com/patents/US7349753.

Pan, J.-N., Li, C.-I., Wu, J.-J., 2016. A new approach to detecting the process changes for multistage systems. Expert Syst. Appl. 62, 293–301. http://dx.doi.org/10.1016/j.eswa.2016.06.037.

Shi, J., 2006. Stream of Variation Modeling and Analysis for Multistage Manufacturing Processes. CRC press, Boca Raton, Fl.

Shi, J., Zhou, S., 2009. Quality control and improvement for multistage systems: a survey. IIE Trans. 41 (9), 744–753. http://dx.doi.org/10.1080/07408170902966344.

Zhang, F., Jiang, P., 2011. Complexity analysis of distributed measuring and sensing network in multistage machining processes. J. Intell. Manuf. 24 (1), 55–69. http://dx.doi.org/10.1007/s10845-011-0538-0.

Zhong, J., Liu, J., Shi, J., 2010. Predictive control considering model uncertainty for variation reduction in multistage assembly processes. IEEE Trans. Autom. Sci. Eng. 7 (4), 724–735. http://dx.doi.org/10.1109/TASE.2009.2038714.

Numerical optimization strategies for springback compensation in sheet metal forming

*A. Maia**, *E. Ferreira**,[†], *M.C. Oliveira*[†], *L.F. Menezes*[†], *A. Andrade-Campos**
*University of Aveiro, Aveiro, Portugal, [†]University of Coimbra, Coimbra, Portugal

3.1 Introduction

The main catalyst to the acceleration of the design stage of new forming processes is the increasing need for new and more complex parts. However, the accuracy requirements are getting stricter. Sheet metal forming is a major industrial process mainly due to its cost efficiency after the establishment of the process design (Tisza, 2014; Chang, 2015). Nevertheless, the process design, including the design of the tools geometry and loading conditions, is not straightforward much as a consequence of the strong nonlinear character of the problem. The emphasis in this issue goes to the springback effect or elastic recovery, which is one of the main causes of inaccuracy between the desired geometry and the actual obtained part. The level of this deviation induces a serious quality problem on production, requiring some compensation or control strategy.

Although there is already commercial software that can predict the shape of a manufactured part given the initial and boundary conditions (direct problem), and iteratively adjust these conditions to achieve the desired part (inverse problem), its efficiency is still not optimal. Additionally, the challenge of compensating the nonlinear springback effect is not efficiently answered by the traditional "trial-and-error" process, which leads to active research in this field.

Nowadays, there is no consensus considering the most effective strategy for springback compensation in sheet metal forming. Therefore, the main goal of this work is to discuss different evaluation, optimization, and parameterization strategies, in order to compensate springback and contribute to efficient solutions.

Several types of geometry parameterizations can be found in the literature for modeling the sheet metal forming tools. Examples are the classical geometric parameterization, which resorts to geometrical dimensioning (Dimas and Briassoulis, 1999; Teixeira et al., 2012), and more advanced parameterizations such as B-splines, T-splines, and NURBS (Lingbeek, 2003; Sederberg et al., 2004; Ponthot and Kleinermann, 2006). These parameterization strategies can be used not only to model the tool but to define the optimization variables in a design process. These design variables can be updated by the optimization algorithms in order to try to achieve a certain goal. Whatever the optimization algorithm adopted, its

Computational Methods and Production Engineering. http://dx.doi.org/10.1016/B978-0-85709-481-0.00003-3

progress and success is evaluated through an objective function that measures the gap between the current solution and the desired final geometry. This objective function can be estimated based on either experimental or numerical results. The last, employing Finite Element Analysis (FEA), can be performed using implicit or explicit time integration methods.

Considering the modeling approaches of the addressed problem, the more recurrent strategies in literature are the Finite Element Model Updating (FEMU) (Steenackers et al., 2007) and Metamodeling Optimization (Wiebenga et al., 2012) strategies. Whereas the former resorts to the sequential use of simulations and evaluations in each iteration of the optimization algorithm, the latter runs all the evaluations prior to the optimization process, in order to create a metamodel (e.g., Response Surface Methods [RSM]) (Khuri and Mukhopadhyay, 2010), which is then used in the optimization process. Nevertheless, both methods resort to the use of optimization algorithms to find the best solution to compensate the geometrical gap between the attained and the target geometries.

In this work, both these strategies are analyzed resorting to a 2D problem: the U-rail benchmark. The experimental and numerical results (obtained using two simulation softwares, with distinct time integration schemes) are then compared and discussed taking into account other results from the literature.

3.2 Springback compensation strategies

A single-stage sheet metal forming process is divided into three phases: (i) the initial positioning of the tools; (ii) the forming by the tools; and (iii) the removal of the tools with subsequent springback. Springback or elastic recover is an undesired side effect of sheet metal forming, which is proportional to the ratio between the residual stresses and the Young's modulus (Eggertsen and Mattiasson, 2009). This relation leads to a first approach to minimize this side effect: the increase of the plastic deformation by incrementing the Blank Holder Force (BHF) and/or other restraining forces (Gösling et al., 2011; Liao et al., 2013; Lokhande and Nandedkar, 2014). However, this approach may induce serious quality problems in the final part, as it leads to a reduction of the material flow during the forming phase. This may generate local thinning of the blank and consequent fracture (Gösling et al., 2011). In order to minimize this effect, some strategies based on the variation of the BHF along the whole process have been researched with promising results. One example is the Intermediate Restraining (Liu et al., 2002), where the force is applied in two phases: an initial phase with only enough pressure to prevent wrinkling, though letting the blank flow between the tools, and a second one, with more intense forces, to induce plastic deformation.

Other possible approaches are Springback Compensation methodologies. These methodologies acknowledge the springback as unavoidable. Thus, instead of trying to reduce it, they counterbalance the springback so that the final part has the desired shape.

One springback compensation strategy is the Force Descriptor Method (FDM) (Karafillis and Boyce, 1996). This method computes the distribution of the traction forces on the workpiece when fully loaded. Then, this information is used to estimate the spring forward of the desired shape, based on FEA, thus obtaining the part that after springback has the reference shape. However, it is not very useful in practical examples, as it presents severe restrictions such as the requirement of symmetry of the part and convergence up to 3% of the geometrical errors.

Another approach is through the Displacement Adjustment (DA) method (Gan and Wagoner, 2004), which consists on the displacement of the tool's surface on the opposite direction of the springback, done in just one step (Meinders et al., 2008; Yang and Ruan, 2011; Liao et al., 2013) or iteratively (Lingbeek et al., 2006). The later concept has revealed itself more effective in areas strongly affected by plastic deformation. In spite of its good practical results, this methodology has some drawbacks such as the difficulty in aligning the CAD model and the forming piece or the tendency of the compensated surface to become rougher (Liao et al., 2013). The former problem may be solved by the Smooth Displacement Adjustment (SDA) method. This methodology approximates the geometric error by smooth functions, using boundary constraints if restrictions are needed (Meinders et al., 2008). Other improvement on the DA strategy is the Comprehensive Compensation method, where instead of adopting the opposite direction of the springback, the best direction and magnitude of the adjustment is computed. The iterative application of this method has achieved promising results (Yang and Ruan, 2011).

Another possible evolution of the DA method, introduced by Lingbeek (2003) and Lingbeek et al. (2005), is the Control Surface Method. This approach uses flexible surfaces to approximate the part's geometry, subsequently using the DA algorithm to make the required adjustments, to obtain the surface to be used on the redesign of the tools. Even though this methodology achieved good results, including in industrial case studies, it is not possible to achieve perfect compensation of the springback and the algorithm often requires human intervention, not being suitable for an automatic implementation.

The Direct Curvature Method (DCA) is another available strategy. Instead of handling the tool as a whole, it improves different parts of the tool independently through a dynamic compensation factor (Liao et al., 2013). The principle behind it is that for a curvature k, the final curvature, which is equal to the tools curvature plus the one induced by springback, should correspond to the desired geometry, within an acceptable tolerance ϵ.

The computation of the optimal compensation factors and adjustments can be done through a "trial-and-error" methodology using (i) experimental or (ii) numerical trials. Nevertheless, springback prediction by FEM is sensitive to numerical factors. For instance, Xu et al. (2004) concluded that springback is sensitive to the blank sheet discretization, particularly the number of sheet elements in contact with the die radius and the number of integration points. Another example can be found in Gösling et al. (2011) where, besides the tool modifications, stress conditions and work-piece stiffness adjustments are performed in order to obtain the desired final piece. In this study, the numerical parameters were determined in order to minimize their impact in the results. Thus the influence of these factors is not considered in this study, where the design variables considered are only the external BHF and the tools surface geometry.

The previously mentioned methodologies can either be employed through an FEMU strategy or through metamodeling optimization. On the one hand, FEMU uses Finite Element simulations in order to evaluate the difference between the actual data and the reference data, whenever is necessary. Its iterative character requires several sequential evaluations. This strategy can be found in literature applied to springback compensation by authors such as, for example, Steenackers et al. (2007), Ponthot and Kleinermann (2006), and Teixeira et al. (2012).

On the other hand, Robust Optimization and metamodels are based on the attempt to compute the model that best describes the existing data using several statistical techniques (Khuri and Mukhopadhyay, 2010). The most popular metamodels used in metal forming are the RSM. In the scope of this approach, it is very important the selection of the data (obtained experimentally or numerically) through Design of Experiments (DoE) techniques, such as the Taguchi method (Khuri and Mukhopadhyay, 2010) or the Central Composition Design (Montgomery, 2001), among others. After collecting and analyzing the data selected through a DoE, a model is adjusted as a result of the fitting process, which is evaluated statistically in order to be validated. In literature, this is commonly done through an analysis of variance (ANOVA) analysis (Khuri and Mukhopadhyay, 2010; Teixeira et al., 2012). In sheet metal forming, more specifically in springback compensation, this method has been increasingly employed over the last years and includes contributions from several authors, such as Naceur et al. (2005, 2006), Wiebenga et al. (2012), Wiebenga and van den Boogaard (2014), and Teixeira et al. (2012).

Both these strategies resort to optimization algorithms in order to achieve the main objective: springback compensation. Optimization algorithms search for the set of input variable that minimizes or maximizes a cost function. In case of inverse problems, the cost function is the mathematical formulation that measures the deviation between the current state of the solution and the given objective. The aforementioned algorithms may be classified into (i) nature-inspired algorithms (Zang et al., 2010; Kaydani and Mohebbi, 2013), (ii) gradient-based methods (Noel, 2012), (iii) artificial intelligence (AI) (Kaydani and Mohebbi, 2013), and (iv) direct search (Gao and Han, 2012; Rajan and Malakar, 2015). Although the gradient-based algorithms present faster convergence and are fairly efficient, they may be trapped in local minima. However, due to the previously mentioned characteristic, these methods are fairly presented in the literature applied to metal forming and can be found in Teixeira et al. (2012), Bonte (2007), and Carvalho et al. (2008). On the other hand, the nature-inspired algorithms probabilistically converge for the global minimum requiring a large number of evaluations of the cost function, which leads to high computational costs. Applications of these methods can be found in sheet metal forming problems, such as the use of genetic algorithms (Kahhal et al., 2013) or bee colony algorithms (Sun et al., 2012). Similarly, AI algorithms also have the disadvantage of depending strongly on the data set size, as discussed by Forcellese et al. (1998) regarding springback control. Knowing that the best results require massive quantities of data for their training, this option is often considered cost-inefficient. However, when well calibrated, the AI algorithms provide very accurate results as concluded by Kazan et al. (2009). The direct search algorithms do not need any prior information about the optimization problem, since they explore the area around the starting point and evolve in the most favorable direction. These algorithms have the advantage of dealing with success with nondifferentiable or noncontinuous problems and can be used in non-a priori problems, as is the case of springback compensation (Lingbeek, 2003).

All springback compensation strategies have advantages and disadvantages, being the selection strongly dependent on the specific application and on the amount of available data. There is also the possibility of combining several approaches in order to take advantage of their best features (Ponthot and Kleinermann, 2006; Kahhal et al., 2013; Serban et al., 2013). All of these concepts are schematically illustrated in Fig. 3.1. Considering the classification defined in Fig. 3.1, the work developed in this paper focuses on approaches similar to the DCA method, combined with the control of the BHF.

In this chapter, the strategies highlighted in green in Fig. 3.1 are analyzed in different sections. Firstly, a discussion on the modeling approaches adopted for the problem is done in Section 3.3. The FEMU strategy is compared with the metamodeling optimization approach, which comprises several types of metamodels: multiple linear regression (MLR, first-order model), polynomial regression (second-order model), and universal kriging. The methods for evaluating the objective function (the differences between obtained and required geometries) are examined in Section 3.4, namely the FEA and the Experimental procedures. The former is explored considering both explicit and implicit time integration and springback prediction using two different strategies: tools reversed displacement and one single step.

The optimization algorithms employed are reviewed in Section 3.5, particularly the gradient-based and direct search algorithms, which are used in this chapter. The parameterization methodologies, namely the classical and the NURBS parameterizations, are compared and discussed in Section 3.6.

The previously mentioned strategies (modeling, evaluation, optimization, and parameterization) are applied to the U-rail benchmark. The description and justification for the choice of this case study are presented in Section 3.7. The results achieved are then compared and analyzed in Section 3.8, considering the computational effort and the quality of the final pieces. Finally, the main conclusions are drawn in Section 3.9.

3.3 Modeling strategies

3.3.1 Metamodeling optimization: Response surface methodology

The RSM belongs to the experimental statistical field. The aim of this methodology is to explore the relationship between one or more response variables and a set of factors or independent variables, which are then characterized by an equation (model), suitable to be used for optimization. In order to reduce the number of tests, a DoE is used to retrieve the most relevant data with the least cost possible (Wiebenga and van den Boogaard, 2014). The dimension of any dataset obtained through a DoE could be

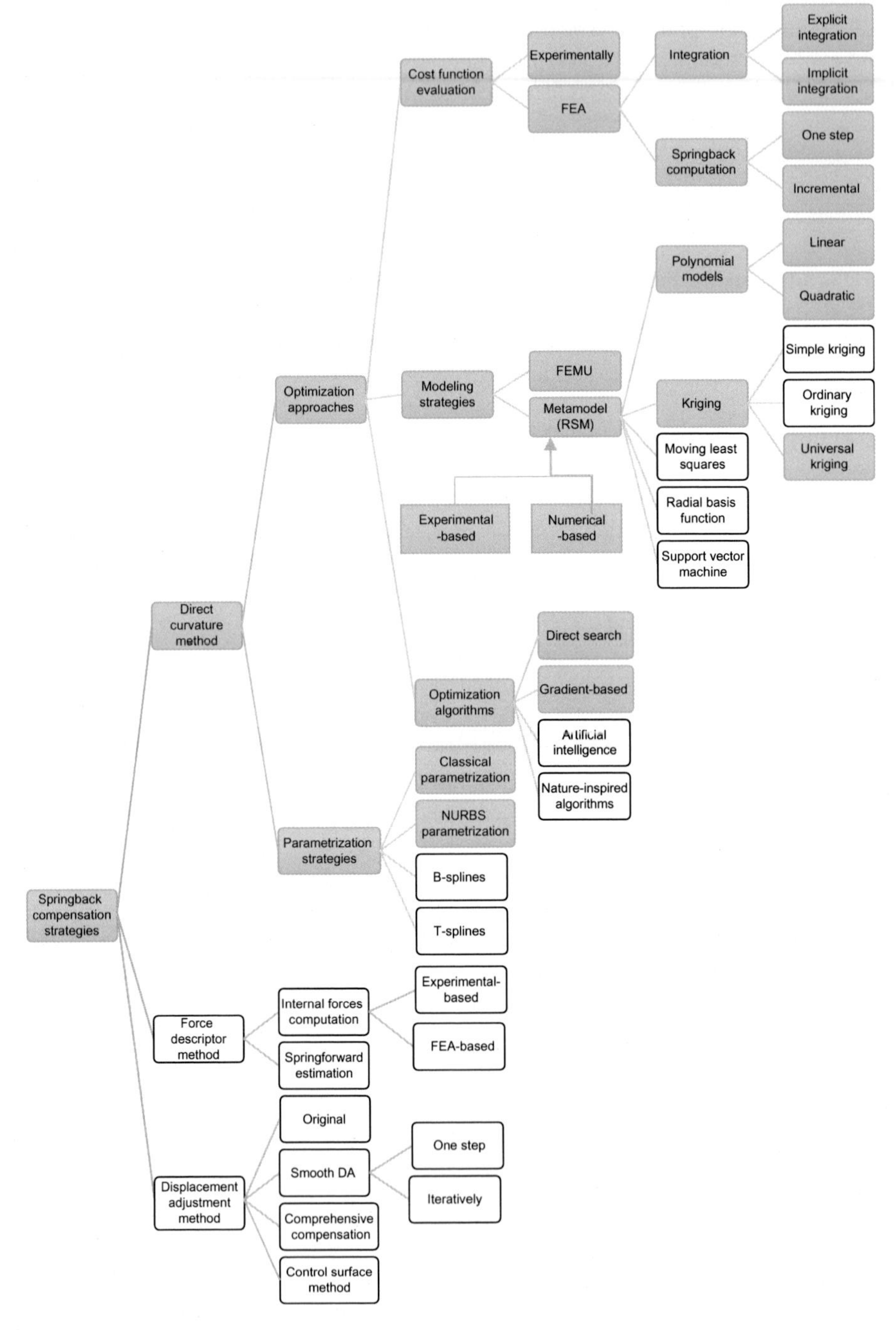

Fig. 3.1 Springback compensation strategies. The strategies implemented in this work are highlighted in *green*.

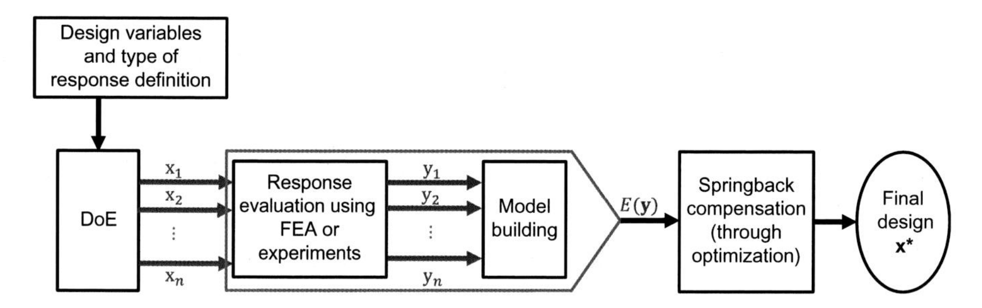

Fig. 3.2 Process schematics for metamodeling optimization using a Response Surface Methodology.

expressed as $n \times (p + k)$, where n is the number of observations, p the number of independent variables, and k the number of response variables.

The application of RSM in a design process is summarized in Fig. 3.2. The first stage of metamodeling consists on defining the goal, the design variables or controllable factors, the responses, and also the validation measures. Then, in order to reduce computational time, DoE strategies can be used. The following stage is the more time expensive due to the execution of forming simulations or experimental tests in order to obtain the dataset of response measurements for all design variables. Finally, the metamodel is fitted, validated, and analyzed. Several metamodels are usually applied, such as multiple linear and polynomial regression, universal kriging, etc.

3.3.1.1 First-order model: MLR

In this work, **y** is defined as the vector of intermediate response models that enable the evaluation of the springback effect and **x** is the vector of the variables used to control the springback effect. Therefore, the response variable **y** is modeled by a set of p independent controllable factors $x_1, \ldots, x_p$ expressed as (Draper and Smith, 2014)

$$y_i = \beta_0 + \beta_1 x_{i1} + \cdots + \beta_p x_{ip} + \epsilon_i, \quad i = 1, \ldots, n, \tag{3.1}$$

where $\epsilon_i \sim N(0, \sigma^2)$ and n is the number of observations. β_0 is the Y-intercept of the regression surface and each β_j, with $j = 1, \ldots, p$, represents the slope of the regression surface regarding its associated variable x_j. All coefficient parameters are estimated by a least-squares method, mathematically defined by

$$\hat{\boldsymbol{\beta}} = (\mathbf{X}^T\mathbf{X})^{-1}\mathbf{X}^T\mathbf{y}, \tag{3.2}$$

$$\hat{\mathbf{y}} = \mathbf{X}(\mathbf{X}^T\mathbf{X})^{-1}\mathbf{X}^T\mathbf{y}, \tag{3.3}$$

$$\mathbf{e} = \mathbf{y} - \hat{\mathbf{y}}, \tag{3.4}$$

where $\hat{\boldsymbol{\beta}}$ is the vector of the estimated coefficients, $\hat{\mathbf{y}}$ contains the predicted values, $\mathbf{X}$ is an $n \times p$ matrix, and $\mathbf{e}$ is the vector of the errors.

In order to analyze if the coefficient obtained for a controllable factor is significant, the t-student test is performed. In statistical hypothesis testing, the hypothesis to be tested is called the null hypothesis H_0, against an alternative hypothesis H_1. For this t-student test, H_0 considers the coefficient β_j as zero (nonsignificant) and it is expressed by H_0: $\beta_j = 0$. The alternative hypothesis considers the coefficient β_j nonzero (significant) and it is expressed by H_1: $\beta_j \neq 0$. Therefore, the test to be performed is expressed by

$$H_0: \beta_j = 0 \text{ versus } H_1: \beta_j \neq 0, \text{ for } j = 1, \ldots, p. \tag{3.5}$$

Therefore, for a significance level α, the null hypothesis H_0 is rejected when the P-value $< \alpha$, meaning that the controllable factor x_j is significant (Fox, 1997; Orme and Combs-Orme, 2009). Moreover, in order to reinforce the conclusion, β_j is only included in the model if its P-value is lower than all the following significant levels: $\alpha = 0.10$, $\alpha = 0.05$, and $\alpha = 0.01$.

Afterwards, the ANOVA is performed to validate the model, using the Fisher test. In this test, the null hypothesis states that all coefficients are zero and it is expressed as H_0: $\beta_1 = \cdots = \beta_p = 0$. The alternative hypothesis means that at least one coefficient is nonzero and it is expressed as H_1: at least one $\beta_j \neq 0$. Therefore, the test to be performed is expressed by

$$H_0: \beta_1 = \cdots = \beta_p = 0 \text{ versus } H_1: \text{ at least one } \beta_j \neq 0, \text{ for } j = 1, \ldots, p. \tag{3.6}$$

If the P-value $< \alpha$, then the null hypothesis is rejected, meaning that the controllable factors have a significant contribution to explain the variability of the response variable.

Once all intermediate response models are determined, the model that quantifies the springback using the previous selected design variables can also be built. Then, this model can be used as a cost function in order to find the values of the design variables that reduce (compensate) the springback. Therefore, this process is reduced to a design by an optimization procedure that uses the RSM model as objective function.

3.3.1.2 Second-order model: Polynomial regression (P.2)

The polynomial regression model can be described as:

$$y = \beta_0 + \sum_{i=1}^{p} \beta_i x_i + \sum_{i=1}^{p} \beta_{ii} x_i^2 + \sum_{i=1}^{p-1} \sum_{j=2, i<j}^{p} \beta_{ij} x_i x_j + \epsilon, \text{ with } i, j = 1, \ldots, p, \tag{3.7}$$

where $\epsilon \sim N(0, \sigma^2)$ and p is the number of independent controllable factors.

The validation of the significant coefficients and ANOVA is performed as described in Section 3.3.1.1. Furthermore, for this case, the adjusted R-squared (Montgomery, 2001) of the multiple regression can be calculated. When its value is 0.5 or lower, it means that the regression explains 50% or less of the variation data.

3.3.1.3 Universal kriging (U.K.)

Kriging or Gaussian process regression is a commonly used method of interpolation of spatial data (e.g., Simpson, 1998). Kriging models are developed with a deterministic surface $f(\mathbf{x})$ and a spatially autocorrelated error $Z(\mathbf{x})$ (Simpson, 1998). A kriging model can be given by

$$\mathbf{y} = f(\mathbf{x}) + Z(\mathbf{x}). \tag{3.8}$$

For universal kriging, $f(\mathbf{x})$ is a known polynomial trend model and $Z(\mathbf{x})$ is assumed to be a Gaussian stochastic process with mean equal to zero, variance σ^2 and nonzero covariance. The covariance matrix is given by

$$\mathbf{Cov}\left[Z(x_i), Z(x_j)\right] = \sigma^2 \mathbf{R}(x_i, x_j), \quad \text{with}\ i, j = 1, \ldots, p, \tag{3.9}$$

where $\mathbf{R}(x_i, x_j)$ is the spatial correlation between x_i and x_j and p is the number of independent controllable factors. The correlation function or the semivariogram has to be fitted to quantify how data are related through its distances (Forrester et al., 2008). In this work, exponential variogram is considered and the semivariogram is built under the assumption of intrinsic stationarity. The coefficients are fitted as in Cressie (1991):

$$\hat{\beta} = \left(\mathbf{X}^T \mathbf{Cov}^{-1} \mathbf{X}\right)^{-1} \mathbf{X}^T \mathbf{Cov}^{-1} \mathbf{y}. \tag{3.10}$$

ANOVA is also performed for kriging regression in order to validate its results.

3.3.2 Finite element model updating

Another possible modeling approach is the FEMU (e.g., Steenackers et al., 2007), where the optimization algorithm resorts to one simulation every time a set of design variables is evaluated, as shown in Fig. 3.3.

In this process, an initial evaluation of the geometrical deviation is performed using FEA for a preestablished initial design. If the stopping criteria are not met, the previously computed deviation and design are taken as input in the optimization algorithm, which suggests a new design. This design is in turn simulated and evaluated, closing a cycle which is repeated iteratively until the stopping criteria are verified. The stopping

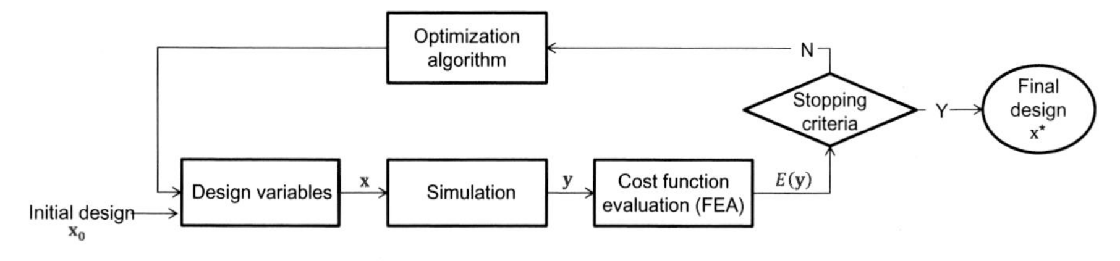

Fig. 3.3 Process schematics for Finite Element Model Updating.

criteria generally include the maximum number of iterations and a threshold, which is compared with the value of the cost function.

This approach is very reliable in the sense that it is not based on interpolations of the springback behavior. Instead, it computes the actual values of the measurable variables for each set of design variables. However, it can be computationally expensive and sensitive to numerical noise, as each iteration of the optimization algorithm requires at least one new simulation and the evaluation of the geometric deviation between the iteration's result and the reference (as cost function).

3.4 Evaluation strategies

3.4.1 Cost function formulation

There are several approaches on how the geometrical deviation between the obtained and the reference parts should be measured (Yang and Ruan, 2011). From the Euclidean 3D distance between corresponding pairs of nodes of the formed blank meshes to the distance between those nodes along one well-defined direction (e.g., X-axis), or variants of these, the choice of the approach to be emploied is strongly related to the case study. However, the use of distances often requires a pairing algorithm between the nodes on the designed and reference parts, which can be difficult and inefficient.

More often, the option of comparing a set of deduced geometric dimensions (similar to the ones used in classical geometrical parameterization) (Naceur et al., 2006; Teixeira et al., 2012; Wiebenga and van den Boogaard, 2014) is enough for the purposes of the work and more efficient, due to the reduced number of intermediate values to be computed and to its ability to be adapted to the experimental parts. This method establishes a set of dimensions, such as angles or curvatures, to be measured and compares the difference between the reference part values and the ones predicted or experimentally obtained. These values are then used as input in a cost function, which computes the final deviation between the parts. In this work, the cost function E is defined as

$$E(\mathbf{y}) = \sum_{i=1}^{n} b_i (y_i - y_i^{\mathrm{ref}})^2, \quad \text{with } i = 1, \ldots, n_{\mathrm{obs}}, \tag{3.11}$$

where $y_i(\mathbf{x})$ and y_i^{ref} are the observed and the desired values of the intermediate responses at location i (see Section 3.3) and b_i is the weight associated to the ith parameter. The weight is also used as a scale factor in order to level the differences of magnitude orders. The evaluation of $y_i(\mathbf{x})$ requires an FEA simulation or performing an experimental test.

3.4.2 FEA software

A problem inherent to the use of the Finite Element Method (FEM) is the numerical noise induced by round-off errors, mesh discretization and instability of contact conditions (Wiebenga and van den Boogaard, 2014). This noise may lead to variations in the number of increments of the simulations and, consequently, to different results and evolution of the design optimization process. In extreme cases, it can also create local minima, trapping the optimization process and leading to inaccurate results. This fact builds up the necessity of accounting for the noise and employing strategies to overcome it, such as multistart strategies.

In this work, two different time-integration methodologies are considered: implicit and hybrid (explicit for the forming and implicit for the springback step). The finite element implicit integration is implemented through DD3IMP (Menezes and Teodosiu, 2000), a software specifically developed to simulate sheet metal forming processes, which uses an updated Lagrangian formulation and a predict-correction scheme to determine the equilibrium state (Oliveira et al., 2007). The tools are assumed as rigid and handled by means of Nagata patches (Neto et al., 2014). This software requires the use of external programs to handle the preprocessing of the input files, particularly the IGES files that contain the NURBS surfaces used to describe the tools.

On the other hand, the simulations run through Abaqus (Hibbitt et al., 2011) have hybrid integration and are carried out in two stages. The first stage, that performs the forming stage, is executed using Abaqus/Explicit with dynamic explicit formulation, which employs an explicit central-difference time integration rule, that in turn allows performing a large number of small time increments efficiently. Then, for the scope of springback computation, a script in Python is used to parameterize the geometry and compute the objective function. In this second stage the implicit Abaqus/Standard is used incrementally (Hibbitt et al., 2011). The preprocessing and adjustment of the simulation tools is, in this case, done internally by Abaqus through external command files given by the user.

The numerical results given by the simulations' outputs, for both time-integration methodologies, are then evaluated by the same external software, which measures the deviation between the final part produced and the reference one.

3.4.3 Experimental analysis

The validation of the numerical results is a crucial component of simulation work, particularly when combined with optimization strategies. It is important that the numerical procedures employed give results as close as possible to the experimental ones, in order to be useful.

The experimental component of this works was done using a mild steel (USB steel). The contact surfaces of this metal sheet were lubricated homogeneously within thin tolerances, as fully detailed in Santos et al. (2004). For this experimental process, the BHF is controlled through nitrogen gas springs, which allow the linear increase of the force with the punch displacement. The remaining details of these procedures can be seen in Santos et al. (2004, 2005) and Teixeira (2005).

The collect data (Teixeira et al., 2012) were used to validate the simulation software employed through a benchmark (see Section 3.7). This experimental data were also used to build a metamodel which was compared with the one obtained by FEA (Section 3.8).

3.5 Optimization algorithms

3.5.1 Generalized reduced gradient

The Generalized Reduced Gradient (GRG) Method proposed by Lasdon et al. (1978) is one of the most popular methods to solve problems of nonlinear optimization (Chapra and Canale, 2009), requiring only that the objective function is differentiable. The main idea of this method is to solve the nonlinear problem dealing with active inequalities. The variables are separated into a set of basic (dependent) variables and nonbasic (independent) variables. Then, the reduced gradient is computed in order to find the minimum in the search direction. This process is repeated until the convergence is obtained (Venkataraman, 2009).

3.5.2 Direct search algorithm

The Nelder-Mead (N-M) Simplex algorithm (Lingbeek et al., 2005; Nocedal and Wright, 2006; Rajan and Malakar, 2015) is a nonconstraint direct search method, which relies on the construction of a simplex of $N+1$ vertices, for an N-dimensional problem. Then, it iteratively replaces its vertices for new ones with lower values of the cost function. Its main advantage is its independence to the gradient of the cost function or any approximation, which means that it is applicable to nondifferentiable functions or to cases where the gradient is unknown. Nevertheless, the algorithm needs to evaluate more points, becoming more time-consuming. Another disadvantage is that near local minima, the algorithm may enter in oscillation, not converging to a single value.

3.5.3 Least-squares gradient-based algorithm

The Levenberg-Marquardt (L-M) (Teixeira, 2005; Bonte, 2007; Carvalho et al., 2008) is a gradient-based method, similar to the Newton least-square method, having, however, a stabilization parameter, μ_k, in order to improve the algorithm's behavior around minima.

The major difficulty of this algorithm for the springback compensation is its sensitivity to noise. The noise does not only affect the evaluations of the cost function (possibly even introducing extra local minima), but also the evaluations necessary to compute the Jacobian. This last effect may lead the algorithm to evolve in wrong directions. Another possible complication regarding this algorithm is its requirement, shared by all gradient-based algorithms, that the number of variables should be equal or fewer than the number of observations. For complex geometries, with many degrees of freedom (dof), this limitation should be taken into account.

3.6 Parameterization strategies

There are several types of parameterizations (Piegl and Tiller, 1997; Dimas and Briassoulis, 1999; Ponthot and Kleinermann, 2006), being the oldest one the classical geometric parameterization. This type of representation resorts to geometrical dimensioning, such as lengths, radii, and/or angles, to describe a shape (Ponthot and Kleinermann, 2006). Though being more intuitive to the human user and commonly used to characterize project designs, this type of parameterization does not offer flexibility. Furthermore, when employed in very complex geometries it resorts to approximations, discarding the representation of details, or increases the number of design variables, with the risk of becoming computationally expensive and complex. Nevertheless, this parameterization is widely used in industry, mainly in tool manufacturing due to its intuitiveness and tradition.

However, more advanced parameterizations such as the T-splines (Sederberg et al., 2004) are available. Though presenting important advantages, namely the reduced number of control points, which lead to their superior computational efficiency without compromising the representation accuracy or resorting to simplifications, this parameterization is still not spread in the industry and it is only applied in some in-use software.

B-splines (Piegl and Tiller, 1997) is an intermediate option, being more used in CAD software than T-splines and more flexible than classical parameterization. This technique allows representing accurately complex shapes, using fewer variables. A more used and powerful extension of these parameterizations is the Non-Uniform Rational B-Splines (NURBS). The use of NURBS (Piegl and Tiller, 1997; Dimas and Briassoulis, 1999) as parameterization presents a good balance between accuracy and computational efficiency. Their flexibility and accuracy on representing complex shapes have led to their adoption and standardization at an industrial level, being used on a wide range of applications and software. When they are used to define the design variables for optimization or project design, the NURBS is controlled through their control polygon coordinates and weights (Piegl and Tiller, 1997). An important fact is that industrial standard files, such as the IGES, already have codification particularly adapted to this type of parameterization.

3.7 Case study: U-rail

The U-rail (Liu et al., 2002; Xu et al., 2004; Naceur et al., 2006; Teixeira et al., 2012) is a widely known and used benchmark, being one of the benchmarks of NUMISHEET'93 (Makinouchi et al., 1993). Its 2D profile and longitudinal symmetry, as shown in Fig. 3.4A, allow reducing the computational times by adopting plane strain conditions along the Oy direction (see Fig. 3.4B). Furthermore, the stress gradient along the thickness of the blank makes this benchmark highly influenced by springback, turning it in a reference for the study and comparison of springback compensation techniques.

As shown in Fig. 3.4, the U-rail is composed by a horizontal planar bottom, two upper wings desirably planar and parallel to the bottom, and two vertical walls orthogonal to both surfaces previously mentioned and equally planar. The big challenge in this case is not only to compensate the extra angular opening caused by springback between the walls and the wings/bottom surface, but also the curvature of the vertical wall. In order to compare the different springback compensation strategies either numerical simulations and/or experimental trials were used.

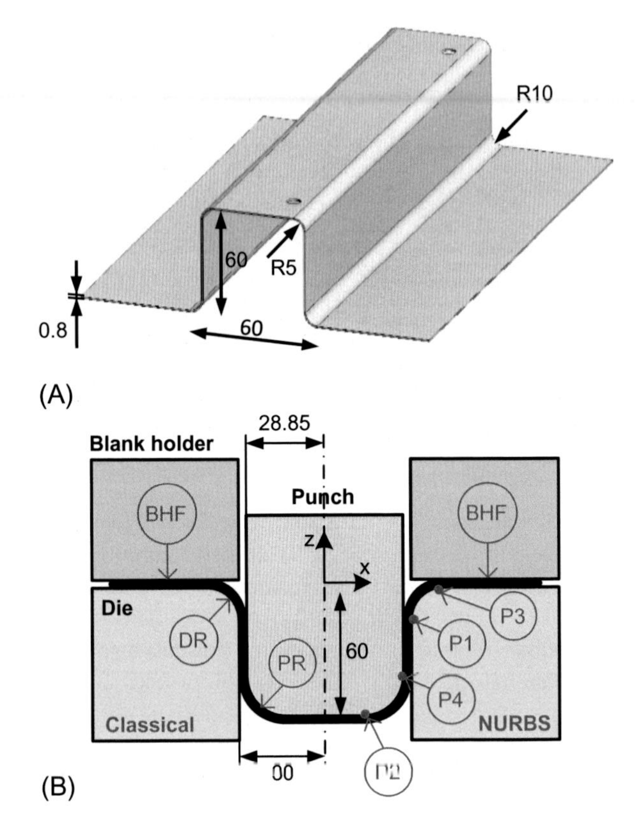

Fig. 3.4 U-rail dimensions and design variables: (A) perspective view and (B) front projection.

The numerical experiments are conducted using the DD3IMP and Abaqus software, as described in Section 3.4.2. In this work, an automatic preprocessor code was developed for DD3IMP. This code works with IGES files and uses the GiD preprocessor to generate the mesh of the blank. The original blank dimension considered is a square of 300 mm.

In DD3IMP, the blank is discretized with eight-node hexahedral solid finite elements, combined with a selective reduced integration technique (Hughes, 1980). The total number of solid elements is $990 \times 1 \times 3 = 2970$ (1 single element in Oy direction and 3 layers of elements through the thickness). On the other hand, in Abaqus, the blank is represented by S4R shell elements with four integration nodes and a reduced integration technique. In this case, and assuming plane strain conditions, a total of 180 shell elements is used. Additionally, for the explicit simulations, a punch velocity of 5 m/s is adopted.

The material used is a mild steel whose behavior can be described by the Swift hardening law. The numerical conditions and material parameters used are listed in Tables 3.1 and 3.2, respectively. The numerical simulation is performed considering an elastoplastic behavior, with isotropic elasticity and anisotropic plasticity (using the Hill'48 yield criterion). The remaining conditions applied to the process are in accordance with values recommended in literature (Xu et al., 2004).

Table 3.1 Conditions for the U-rail test

Numerical conditions	
Tool type	Rigid analytical surface
Friction coefficient	0.15
Process conditions	
Punch travel (mm)	60

Table 3.2 Material properties of the mild steel

Properties	Value
Young's modulus (MPa)	206,629
Poisson's ratio	0.298
Initial yield stress (MPa)	175
K, ε_0, and n from $\sigma = K(\varepsilon_0 + \overline{\varepsilon^p})^n$	488.35, 0.24, 0.0146
r_0, r_{45}, r_{90}	2.09, 1.56, 2.72

It is possible to see in Fig. 3.4B the correspondence used for converting the classical parameterization into NURBS. The classical geometric parameterization, presented in red, resorts to distances and radii (PR and DR), whereas in blue, the design variables are the coordinates of the NURBS control points ($P1$, $P2$, $P3$, and $P4$). On the latter parameterization, some constraints are added in order to guarantee the orthogonality between vertical and horizontal surfaces of the tools or that the concordance between those surfaces is described by a radius. The later constraints can be written as

$$P1_z = (30 - P3_x) = -DR; \quad P4_z = (28.85 - P2_x) = PR, \tag{3.12}$$

where the subscript indicates the coordinate of the point.

The design variables for the classical parameterization are then $\mathrm{x} = [\mathrm{DR} \quad \mathrm{PR} \quad \mathrm{BHF}]^T$ where DR and PR are the radius of the die and punch, respectively, and BHF is the value selected for the blank holder force. When adapting to NURBS parameterization, the design variables are $\mathrm{x} = [P1_z \quad P2_x \quad \mathrm{BHF}]^T$ where $P1_z$ is the coordinate z of point $P1$ of the die and $P2_x$ is the coordinate x of point $P2$ of the punch.

In cases where it is necessary to add constraints to the search space of these parameters, the input space considered in Teixeira et al. (2012) is adopted. The constraint to avoid very large values of BHF is especially important as the increase of this variable is strongly related to necking. These constraints are implemented in the optimization algorithms prior to the simulations through variable transformations, that is

$$\hat{x}_i = \hat{x}_i^{\min} + (\hat{x}_i^{\max} - \hat{x}_i^{\min}) \frac{e^{x_i}}{e^{x_i} + e^{-x_i}}, \tag{3.13}$$

where x_i is the unconstrained design variable and $\hat{x}_i$ is the constrained variable subjected to a maximum $\hat{x}_i^{\max}$ and a minimum $\hat{x}_i^{\min}$. e^{x_i} and e^{-x_i} are, respectively, the values of the natural exponential function of x_i and its symmetric.

In order to fully explore the abilities of using NURBS as design variables, another parameterization is taken with more dof. As shown in Fig. 3.5, the extremity of the tools and some coordinates are kept fixed in order to compare this case study with previous U-rail case studies. For instance, by fixing of the coordinate z of point $P2$ (on the punch) or point $P9$ (on the die), the assumption that the horizontal surfaces of the tools must remain planar and horizontal is adopted in this work. Using this approach, each tool would be described by seven dof, six coordinates, and one weight, which corresponds to a total of 14 design variables for the tools geometry. The BHF must be added to these variables as it is a fundamental characteristic of the process. This leads to a design variables set $\mathrm{x} = [x_1 \ \ldots \ x_{14} \ \ \mathrm{BHF}]^T$. It is worth mentioning that, for some of the possible configurations, the representation of the geometry resorting to classical parameterization would be extremely complex to implement. Similarly to what is done with the three dof case, in this approach there are constraints to the range of the coordinate values, in order to keep the tools' surface smooth and feasible. These boundaries are ensured by variable transformations (Eq. 3.13 with the values listed in Table 3.3).

As the main objective of this case study is to ensure the orthogonality and planarity of the surfaces, therefore compensating the springback, instead of computing the Euclidean distance between matching pairs of points of the two parts, the geometrical gap is computed resorting to the values of the angles θ_1 and θ_2, and the distance h (see Fig. 3.6), and their difference to the respective reference values, respectively,

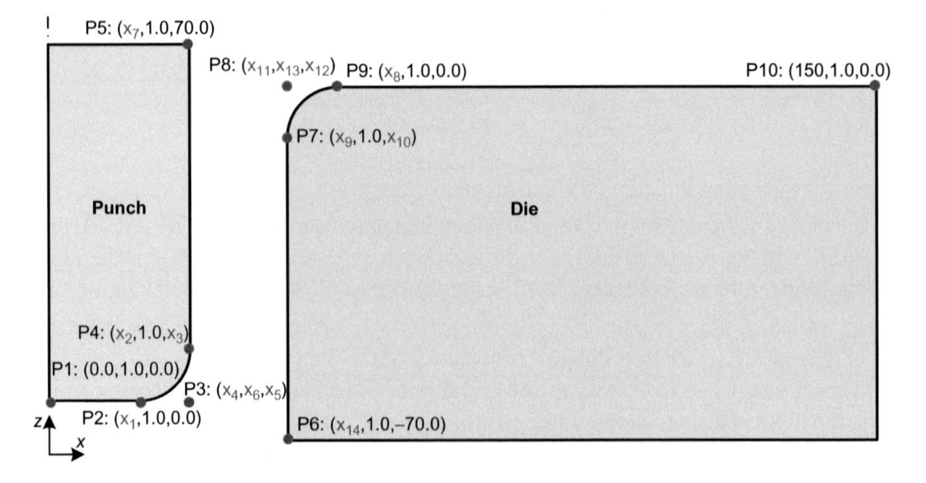

Fig. 3.5 Parameterization of the U-rail tools with 15 dof (e.g., for the plane $y = 1$).

Table 3.3 **Range of the design variables for the design variable set with a total of 15 variables**

Variables	x_1	x_2	x_3	x_4	x_5	x_6	x_7	x_8	x_9	x_{10}	x_{11}	x_{12}	x_{13}	x_{14}	BHF
Max.	28.85	28.85	10.00	x_2	x_3	1.00	28.85	40.00	35.00	−2.00	x_8	0.00	1.00	35.00	300
Min.	18.85	23.85	2.00	x_1	0.00	0.00	23.85	32.00	30.00	−10.00	x_9	x_{10}	0.00	30.00	90

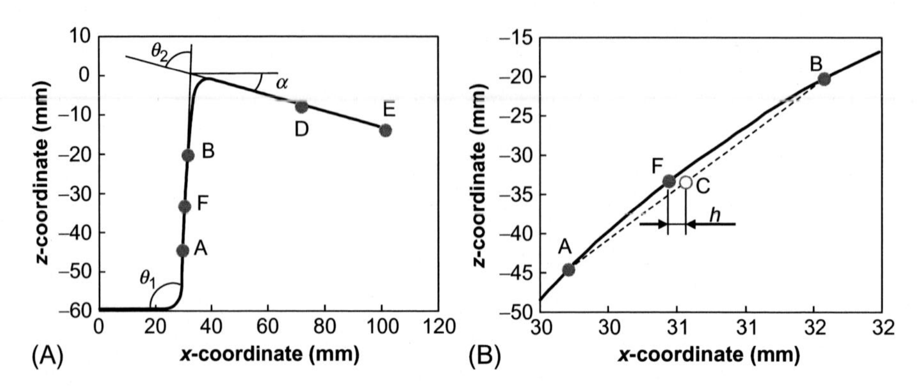

Fig. 3.6 Definition of the evaluation parameters: (A) springback angles and (B) curvature height. Point A (point with z-coordinate $= -45$ mm), point B (point at 25 mm from point A), point C (middle point of line A–B), point D (point at 30 mm from point E), point E (exterior point on the upper wing), and point F (middle point of A–B over the surface).

90 degree, 90 degree, and 0.0 mm. The flange angle α can be computed through angles θ_1 and θ_2 as follows

$$\alpha = \theta_1 - \theta_2. \tag{3.14}$$

Additionally, as the magnitude of the distance h is lower than the one of the angles, this variable is multiplied by 10, so that its influence is not overlooked. Consequently, the cost function can be written as

$$E(\theta_1, \theta_2, h) = (\theta_1 - 90)^2 + (\theta_2 - 90)^2 + (h \times 10)^2, \tag{3.15}$$

where the same weight is adopted for all variables (see Eq. 3.11). The global optimization problem is defined as

$$\begin{aligned} &\text{Minimize} \quad E(\mathbf{y}(\mathbf{x})) \\ &\quad s.t. \quad x_i^{\min} \leq x_i \leq x_i^{\max}, \quad i = 1, \ldots, p, \end{aligned} \tag{3.16}$$

where p is the number of design variables.

Concerning the FEA, the intermediate variables $\mathbf{y}(\mathbf{x})$ are obtained through a post-process code that evaluates the simulations' outputs. In order to verify if the conditions adopted by the FEA accurately capture the real behavior of the sheet metal forming, the results are compared with the experimental ones reported in Teixeira et al. (2012), considering section B as shown in Fig. 3.7. This section was adopted in order to minimize the influence of border effects on the results. The α angle computed by DD3IMP is of 13.66 degree, presenting a deviation of 0.07% to the values predicted by Abaqus, 13.65 degree. Both these values are validated based on the experimental data reported in Teixeira et al. (2012) for the same section of the part, having an error inferior to 2 degree to the experimental results.

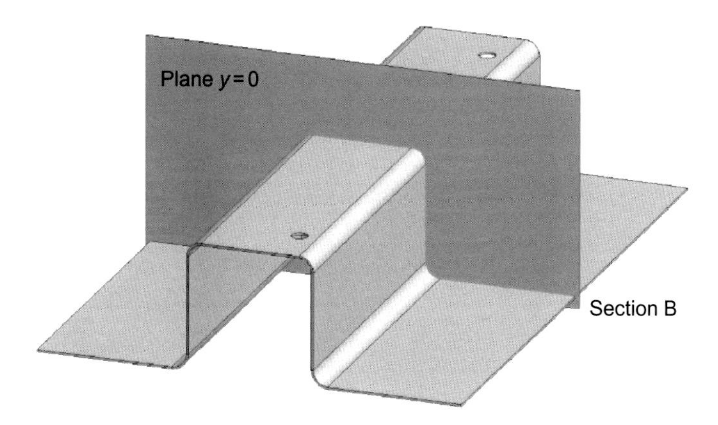

Fig. 3.7 Available sections for evaluation of the experimental parts.
Adapted from Teixeira, P., Andrade-Campos, A., Santos, A., Pires, F., César de Sá, J., 2012. Optimization strategies for springback compensation in sheet metal forming. In: Proceedings of the 1st ECCOMAS Young Investigators Conference—YIC2012, Aveiro.

3.8 Results and discussion

3.8.1 Sensitivity analysis

A sensitivity analysis is performed to evaluate the main effects of all controllable factors in the response variables θ_1, θ_2, and h (see Fig. 3.8). Concerning DD3IMP results (Fig. 3.8A, C, and E) it can be seen that the BHF produces the largest variations in the springback angles and in the curvature height, being most evident when the force is maximum. Other conclusion retrieved from Fig. 3.8C is that the punch radius has a little significance, being the die radius more significant (Fig. 3.8A). In the Abaqus results, the punch radius has no significant effect in θ_1 and θ_2 (Fig. 3.8D), but it influences the vertical wall curvature.

Regarding both simulation softwares, the lower flange angle α is achieved when the die radius is equal to 2.5 mm (Fig. 3.8A and B). As for the BHF, it is also visible that it induces an important variation of the output variables.

Despite having similar trends, it should be noticed that the maximum values attained for DD3IMP results are slightly higher than the ones achieved using Abaqus for all three output variables.

3.8.2 Modeling strategies and evaluation

3.8.2.1 Metamodels

The results hereby presented refer to both experimental and numerical data. The experimental data for the controllable factors used in this work is taken from Teixeira et al. (2012), where the DoE is defined using 3 factors with 3 levels each, which leads to a dataset of 27 experiments. The equations for the MLR response model

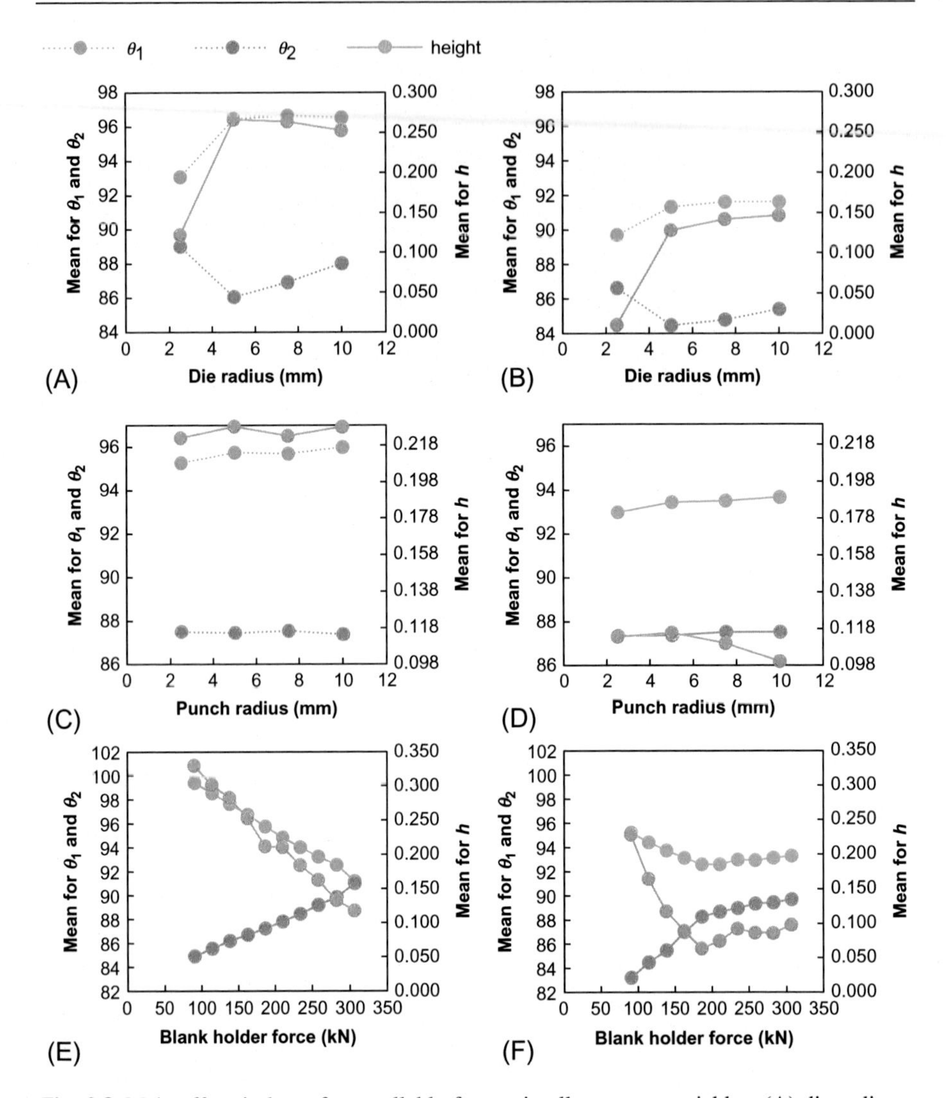

Fig. 3.8 Main effects' plots of controllable factors in all response variables: (A) die radius, (C) punch radius, and (E) Blank Holder Force for DD3IMP simulations; (B) die radius, (D) punch radius, and (F) Blank Holder Force for Abaqus simulations.

are built based on the experimental results collected in section B (see Fig. 3.7) of the part, resulting on the following model

$$\theta_1 = 94.3116 - 0.0121\,\mathrm{BHF} + 0.2543\,\mathrm{DR}, \tag{3.17}$$

$$\theta_2 = 81.8528 + 0.0315\,\mathrm{BHF} - 0.1394\,\mathrm{DR}, \tag{3.18}$$

$$h = 0.1937 - 0.0008\,\mathrm{BHF} + 0.0193\,\mathrm{DR}. \tag{3.19}$$

As the information in this database is insufficient to build second-order models, the MLR is the only model used for experimental data. When minimizing the cost function given by Eq. (3.15) with this model's data, using the GRG algorithm, the achieved solution is

$$\mathbf{x} = [2.5\,\text{mm} \;\; 5.0\,\text{mm} \;\; 287.3\,\text{kN}]^T. \tag{3.20}$$

For a linear model, the multistart approach is not required.

Concerning the numerical data, the total number of simulations ran for both FEA strategies (implicit and hybrid code) is obtained through a DoE of $10 \times 4 \times 4 = 160$. In order to ensure that the data present enough information for the second-order modeling, 10 levels are considered for the BHF[1] and 4 levels are considered for the radii of the die and the punch (2.5, 5, 7.5, and 10 mm). However, the number of completed simulations differs due to the time integration scheme adopted, with DD3IMP finishing successfully (converged) 96% of the simulations and Abaqus 100%. Despite this fact, both datasets contain enough data to fit both first- and second-order models.

For the Abaqus dataset, firstly an MLR is performed in order to identify which controllable factors can significantly influence each response variable. To this end, it is verified that the numerical results agree with the experimental ones: for both the experimental and numerical modelings and for all usual levels of significance, it is concluded that the variable punch radius is not significant. Thus, the MLR response model based on Abaqus dataset is given by

$$\theta_1 = 92.8618 - 0.0075\,\text{BHF} + 0.2477\,\text{DR}, \tag{3.21}$$

$$\theta_2 = 82.1745 + 0.0300\,\text{BHF} - 0.1347\,\text{DR}, \tag{3.22}$$

$$h = 0.1070 - 0.0008\,\text{BHF} + 0.0173\,\text{DR}. \tag{3.23}$$

The same analysis is performed for both second-order models (quadratic regression and universal kriging), considering the same database used for the MLR trial, without the column of the punch radius. The quadratic response model equations can be written as

$$\theta_1 = 96.5444 - 0.0637\,\text{BHF} + 0.9761\,\text{DR} + 0.0001\,\text{BHF}^2 - 0.0659\,\text{DR}^2, \tag{3.24}$$

$$\theta_2 = 78.5031 + 0.1040\,\text{BHF} - 1.2717\,\text{DR} - 0.0002\,\text{BHF}^2 + 0.1119\,\text{DR}^2, \tag{3.25}$$

$$h = 0.2396 - 0.0036\,\text{BHF} + 0.0704\,\text{DR} + 7.5900 \times 10^{-6}\,\text{BHF}^2 - 0.0045\,\text{DR}^2 \tag{3.26}$$

[1] The DoE is designed with the following values for numerical simulations: 90, 114, 138, 162, 186, 210, 234, 258, 282, and 300 kN.

and the universal kriging response model equations are

$$\theta_1 = 96.6389 - 0.0620\,\text{BHF} + 0.8447\,\text{DR} + 0.0001\,\text{BHF}^2 - 0.0557\,\text{DR}^2, \tag{3.27}$$

$$\theta_2 = 78.1065 + 0.1067\,\text{BHF} - 1.2464\,\text{DR} - 0.0002\,\text{BHF}^2 + 0.1123\,\text{DR}^2, \tag{3.28}$$

$$h = 0.2748 - 0.0040\,\text{BHF} + 0.0715\,\text{DR} + 8.2800 \times 10^{-6}\,\text{BHF}^2 - 0.00475\,\text{DR}^2. \tag{3.29}$$

Though several of the previously described optimization algorithms were applied to this model, the optimization results presented refer to the GRG methodology with the multistart approach. The optimum solutions are listed in Table 3.4, where MLR stands for the linear model, P.2 for the quadratic regression, and U.K. for the universal kriging.

Concerning the numerical dataset obtained using DD3IMP, the same procedure is followed. As previously, the MLR is the first metamodel to be fitted, because it can help to identify which controllable factors can significantly influence each response variable. The results obtained are

$$\theta_1 = 99.9556 - 0.0355\,\text{BHF} + 0.3596\,\text{DR}, \tag{3.30}$$

$$\theta_2 = 82.6397 + 0.0260\,\text{BHF} - 0.0402\,\text{DR}, \tag{3.31}$$

$$h = 0.3204 - 0.0009\,\text{BHF} + 0.0142\,\text{DR}. \tag{3.32}$$

Quadratic regression and universal kriging are performed considering the same database as for the MLR case, without accounting the punch radius influence, as previously explained. The quadratic response model equations are given by:

$$\theta_1 = 96.9193 - 0.0477\,\text{BHF} + 2.1376\,\text{DR} + 2.8290 \times 10^{-5}\,\text{BHF}^2 - 0.1455\,\text{DR}^2, \tag{3.33}$$

$$\theta_2 = 84.4513 + 0.0428\,\text{BHF} - 1.4907\,\text{DR} - 0.0029\,\text{BHF} \times \text{DR} + 0.1609\,\text{DR}^2, \tag{3.34}$$

$$h = 0.1836\,\text{DR} - 1.1433 \times 10^{-4}\,\text{BHF} \times \text{DR} - 0.0066\,\text{DR}^2 \tag{3.35}$$

and the universal kriging response model equations are

$$\theta_1 = 95.2935 - 0.0323\,\text{BHF} + 2.2025\,\text{DR} + 7.8555 \times 10^{-6}\,\text{BHF}^2 - 0.1434\,\text{DR}^2, \tag{3.36}$$

$$\theta_2 = 81.8734 + 0.0698\,\text{BHF} - 1.49414\,\text{DR} - 0.0031\,\text{BHF} \times \text{DR} + 0.1597\,\text{DR}^2, \tag{3.37}$$

Table 3.4 Results obtained for the RSM (the optimum solution found using GRG algorithm with multistart) and FEMU (N-M and L-M optimization algorithms) strategies

		Design variables			Evaluation variables					
Strategy	**Methods**	**DR (mm)**	**PR (mm)**	**BHF (kN)**	θ_1 **(°)**	θ_2 **(°)**	***h*** **(mm)**	**Cost funct.**	**Thick. red. (%)**	**No. of simul.**
	Initial	5.00	10.00	90.0	94.37	82.72	0.144	74.144	5.01	1
	Min. Data	2.50	2.50	210.0	90.53	90.00	−0.009	0.287	12.15	160
RSM	MLR	2.50	5.00	276.7	91.09	91.09	−0.001	2.370	14.69	160
	P.2	2.50	5.00	202.0	90.72	89.70	−0.010	0.627	11.75	160
	U.K.	2.50	2.50	201.2	90.72	89.69	−0.014	0.639	11.61	160
FEMU	N-M	2.56	4.77	287.5	90.93	91.14	−0.060	2.535	14.25	62
	L-M	2.52	7.60	223.9	91.40	89.86	0.016	1.998	12.73	18

Note: Cost function evaluation using the Abaqus FEA simulation software.

$$h = 0.1179\,\mathrm{DR} - 1.2747 \times 10^{-4}\,\mathrm{BHF} \times \mathrm{DR} - 0.0064\,\mathrm{DR}^2. \tag{3.38}$$

The optimum solutions are listed in Table 3.5, being also obtained using the GRG algorithm with multistart.

3.8.2.2 FEMU versus metamodels

The numerical results of the implemented approaches are shown in Tables 3.4 and 3.5 according to the FEA strategy employed. In both cases, along with the values of the design (control factors) and evaluation (intermediate) variables, the cost function, the thinning of the blank, and the number of simulations required for each strategy are presented. The minimum value for the cost function found in the dataset used to build the metamodels is also introduced in the row aforementioned as Min. Data. Firstly, considerations concerning the type of approach are made then, the conclusion regarding the most advantageous method for each type of FEA is drawn.

Regarding the metamodels, it is possible to notice that the second-order models (P.2 and U.K.) achieve lower cost function values and less thinning than the first-order ones while using the same number of evaluations. This is expected, because this type of model allows a better fitting of the data, leading therefore to more accurate results.

For the experimental-based first-order model, its coefficients are in the same magnitude order as the numerical ones, being the minimum similar. However, due to the lack of resources, it was not possible to perform a final experiment with the optimized control factors found.

The use of FEMU strategies requires sequential simulations during the optimization process, therefore the choice of the algorithm becomes relevant. Associated with this strategy, the L-M and the N-M algorithms were used. The best results in all cases within the FEMU strategies came from the same initial point $\mathbf{x}_0 = [3.0\text{ mm}, 5.0\text{ mm}, 210.0\text{ kN}]$. Though similar optimal angles were reached, the results are quite different in the distance h. When comparing the computational costs, it is clear that the gradient-based algorithms stand for a significant leverage in the FEMU strategy. The results of the FEMU strategy using the L-M algorithm for both softwares are reached with less than half of the simulations required by the N-M. Therefore, the most advantageous optimization algorithm to associate with an FEMU strategy is a gradient-based due to its fast convergence.

The metamodels approach achieves better results (lower cost function) than the FEMU strategies when using Abaqus, which was not verified for the DD3IMP case. This can be related to the substantially higher values of the distance h for the DD3IMP results, that directly contribute to the cost function values (see Fig. 3.8). Another aspect to consider in this analysis is the computational cost of these two strategies: whereas the metamodels usually require a bigger number of simulations, the FEMU is more unpredictable. Although generally quicker, there is no guarantee about the number of iterations that the algorithms will require to converge, which is also sensitive to the initial solution. Consequently, the number of simulation can vary, especially if the algorithm converges into local minima or starts to oscillate between points of the design space. Additionally, although the metamodels require a large number of simulations,

Table 3.5 Results obtained for the RSM (the optimum solution found using GRG algorithm with multistart) and FEMU (N-M and L-M optimization algorithms) strategies

		Design variables			Evaluation variables					
Strategy	**Methods**	**DR (mm)**	**PR (mm)**	**BHF (kN)**	θ_1 (°)	θ_2 (°)	***h* (mm)**	**Cost funct.**	**Thick. red. (%)**	**No. of simul.**
	Initial	5.00	10.00	90.0	100.34	83.62	0.396	163.286	4.89	1
	Min. Data	2.50	2.50	234.0	90.70	90.42	0.095	1.568	16.19	160
RSM	MLR	2.50	2.50	300.0	89.07	92.61	0.1153	9.000	20.15	160
	P.2	2.50	2.50	259.7	89.75	91.78	0.1380	5.100	16.96	160
	U.K.	2.50	2.50	202.3	92.21	88.92	0.0931	6.900	14.12	160
FEMU	N-M	2.77	1.09	256.0	90.68	90.34	0.024	0.631	17.74	58
	L-M	3.23	1.44	288.2	90.82	90.66	0.090	1.915	15.61	21

Note: Cost function evaluation using the DD3Imp FEA simulation software.

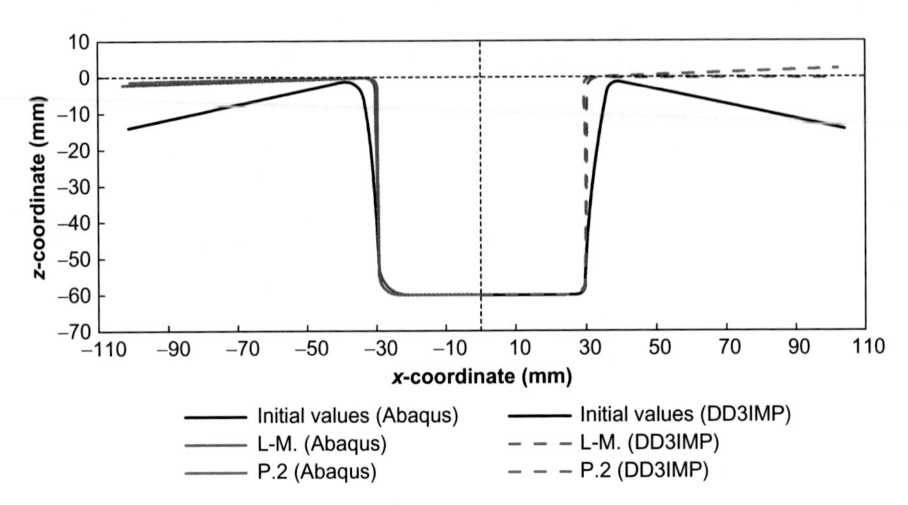

Fig. 3.9 Final parts' geometries for the initial and best FEMU and RSM strategies using Abaqus (*left*) and DD3IMP (*right*) FEA programs.

these are independent and allow a straightforward parallel computation. Therefore, concerning computational time, the advantage cannot be given to the FEMU strategy.

Regarding the Abaqus results, listed in Table 3.4, it is clear that the best result is found in the minimum value of the experiments' database. This is followed by the minimum of the P.2 model and the L-M strategy. In this case, the optimal cost function value achieved by the second-order models is much better than the ones from the FEMU, demonstrating a good fitting of the model to the data. However, when comparing the formed parts' profile (see Fig. 3.9), the observable difference is little. The parts resultant from the P.2 model also have a very reduced level of thinning (less than 12%), indicating that this strategy is the more appropriated to this case study and FEA software.

Concerning the DD3IMP data (see Table 3.5), the best result is given by the N-M algorithm associated with the FEMU strategy, followed by the minimum value of the database and by the L-M. For this case, the best result obtained with the metamodels strategy is still the P.2 model, though its results are far from the ones achieved by the FEMU strategies. Moreover, the attained solution over compensates the springback, as shown in Fig. 3.9. For the FEMU strategies, the number of simulations needed by the N-M algorithm is larger than the one required by the L-M, being the final angles of the part very similar for both cases. Therefore, and also considering the thinning of the blank, which is lower to the later algorithm, the L-M strategy seems to be the most appropriated when using an implicit integration FEA, as the DD3IMP program.

3.8.3 Optimization

When using metamodels, all three optimization algorithms led to very similar results (insignificant differences). Also, as the computations required for this methodology are very simple, there were no significant differences between the algorithms in terms of time, though the gradient-based algorithms reached the minimum with fewer iterations. All the algorithms were tested using the same starting points' set.

Concerning the FEMU strategy, the evolution of each algorithm in the selected case study is shown in Fig. 3.10. Both algorithms used as stopping criterion a tolerance of 10^{-5} for the first-order optimality measure. Concerning the N-M algorithm, standard values for the coefficients were adopted, such as $\alpha = 1.0$, $\gamma = 2.0$, and $\beta = \delta = 0.5$. For the L-M algorithm, the Jacobian matrix is computed at each iteration through a forward finite-difference technique with a perturbation of 2% in the design variable value (these computations are included on the total number of simulation presented in Tables 3.4 and 3.5).

The optimum profiles of the final formed parts are barely distinguishable with very similar cost function values. The largest difference is registered when using the Abaqus FEA software, where each algorithm started from a different starting point. As the starting point of the L-M had a larger value of cost function than the one of the N-M, it seems that the evolution of the former registered a bigger improvement than the one of the latter. However, the final values are rather similar. When considering the computational costs, the gradient-based L-M method is quicker to reach the optimum, taking less than half the number of iterations (and evaluations) and, consequently, half the computation time. On expensive simulations, which is the case, this stands for a great leverage. On the other hand, its strong dependence on the gradient calculation can lead this algorithm to false results in the case of simulations very affected by the noise (see the case of the fourth iteration for the Abaqus case where the cost function for L-M increases). Therefore, the use of a larger perturbation step for the finite-difference gradient calculation (5% of the variable value) than the one typically recommended on literature was employed as a mean to overcome this limitation. The results obtained are consistent with both implicit and hybrid-based codes, pointing to the supremacy of the gradient-based methods in this methodology and case study.

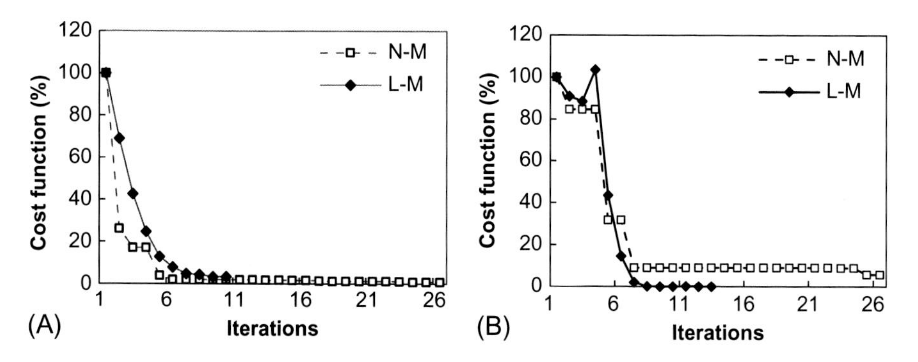

Fig. 3.10 Evolution of the normalized cost function during the optimization processes using (A) DD3IMP and (B) Abaqus FEA programs.

3.8.4 Parameterization

The cases with classical parameterization (3 dof) and NURBS (15 dof) were compared using the N-M algorithm associated with an FEMU strategy, and using the DD3IMP FEA program with IGES files as input. The feature of using IGES files made the whole process of updating the input files easier, as it did not require the transformation of NURBS into classical parameterization. The classical parameterization was ensured by the use of constraints that guaranteed perfectly circular radii. Both cases were carried out using the same set of starting points in the optimization process.

The L-M algorithm was discarded due to the large number of design variables for the NURBS parameterization (15 > 3), which is larger than the number of observation of the manufactured part (see Section 3.7). In the case of metamodels, even using DoE to reduce the number of simulations to perform, the computation time is excessive for the industry. Therefore, the metamodel strategy was also ruled out.

Comparing the evolution of the optimization process with 3 and 15 dof shown in Fig. 3.11, the case with more dof reveals a faster convergence. This is expected due to the increase in the tools flexibility.

Another effect of increasing the number of design variables is the establishment of more local minima. This may induce convergence problems, leading to final solutions that are not the global minimum.

Another interesting point is that, even though the extra dof lend more flexibility to the tools geometry, the convergence tends to a shape very close to the one attained with three design variables, even when evolving from distinct starting points.

An additional case was analyzed. This case used NURBS and, as starting point, the optimum geometry found with the 3 dof. Though the algorithm ran for more than 20 iterations, no improvement is registered. From this analysis, it is concluded that, for this case study, the classical geometric parameterization reaches the best results, though being more time consuming.

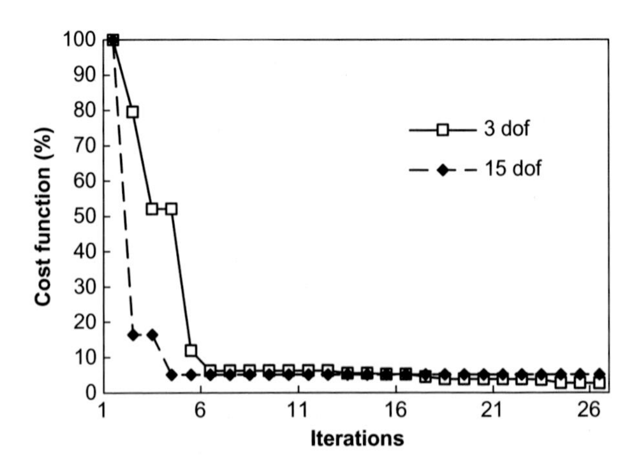

Fig. 3.11 Evolution of the optimization process using 3 and 15 dof.

3.9 Conclusions

In this work, the U-rail benchmark is used to compare different optimization strategies and tools parameterization approaches, with the goal of finding efficient strategies for springback compensation. Regarding the tools parameterization, classical geometric parameterization and NURBS were compared in order to analyze the benefits brought by the flexibility and the required computational time. Concerning the subject of design optimization, the search for the design variables is performed through meta-model RSM optimization and through FEMU strategies.

From the results achieved, it is possible to conclude that both strategies lead to significant improvements over the initial parts and achieve results close to the desired reference. Furthermore, the numerical results are in agreement with the experimental ones, being the trend to reduce the DR and increase the BHF. However, it is also observable that the best conjugation of design variables is FEA's software dependent. This is justified by the different time-integration natures of the software (implicit time integration on DD3IMP vs. hybrid integration on Abaqus).

Regarding RSM, the quadratic model presents the best results. However, the use of RSM strategies requires the achievement of a computationally costly database. When considering these costs, the optimization algorithms, particularly the gradient-based, provide a good alternative, reaching similar values with much fewer simulations.

Regarding the type of parameterization used for the tool's surface, it is possible to conclude that with NURBS, the results are as good as the ones obtained using the geometric parameterization. However, for this case study, the increment of the number of dof does not add extra improvement to the springback compensation process, leading to the geometries obtained with the simple three design variables.

In sum, the results of metamodeling optimization could overcome the ones obtained through FEMU, depending on the simulation software. Moreover, the use of NURBS did not add extra advantage to this case study, although it enabled larger flexibility to the design optimization problem.

Acknowledgments

This work was co-financed by the Portuguese Foundation for Science and Technology via project PTDC/EME-TME/118420/2010 and by the FEDER via the "Programa Operacional Factores de Competividade" of QREN with COMPETE reference: FCOMP-01-0124-FEDER-020465.

References

Bonte, M., 2007. Optimisation strategies for metal forming processes. PhD Thesis, Universiteit Twente.

Carvalho, J.F., Cruz, P.S., Valente, R.A.F., 2008. An integrated methodology for parameter identification and shape optimization in metal forming and structural applications. In: Proceedings of International Conference on Engineering Optimization, Rio de Janeiro.

Chang, K.H., 2015. e-Design: Computer-Aided Engineering Design. Academic Press, London.

Chapra, S., Canale, R., 2009. Numerical Methods for Engineers, sixth ed. McGraw-Hill, New York, NY.

Cressie, N., 1991. Statistics for Spatial Data. New York, NY.

Dimas, E., Briassoulis, D., 1999. 3D geometric modelling based on NURBS: a review. Adv. Eng. Softw. 30 (9–11), 741–751.

Draper, N., Smith, H., 2014. Applied Regression Analysis. New York, NY.

Eggertsen, P.A., Mattiasson, K., 2009. On the modelling of the bending—unbending behaviour for accurate springback predictions. Int. J. Mech. Sci. 51 (7), 547–563.

Forcellese, A., Gabrielli, F., Ruffini, R., 1998. Effect of the training set size on springback control by neural network in an air bending process. J. Mater. Process. Technol. 80–81, 493–500.

Forrester, A., Sobester, A., Keane, A., 2008. Engineering Design via Surrogate Modelling: A Practical Guide. John Wiley & Sons, New York, NY.

Fox, J., 1997. Applied Regression Analysis, Linear Models, and Related Methods. Sage Publications, Inc, Newbury Park, CA.

Gan, W., Wagoner, R., 2004. Die design method for sheet springback. Int. J. Mech. Sci. 46 (7), 1097–1113.

Gao, F., Han, L., 2012. Implementing the Nelder-Mead simplex algorithm with adaptive parameters. Comput. Optim. Appl. 51 (1), 259–277.

Gösling, M., Kracker, H., Brosius, A., Kuhnt, S., Tekkaya, A., 2011. Strategies for springback compensation regarding process robustness. Prod. Eng. 5 (1), 49–57.

Hibbitt, H., Karlsson, B., Sorensen, P., 2011. Abaqus Analysis User's Manual Version 6.10. Dassault Systèmes Simulia Corp., Providence, RI.

Hughes, T.J.R., 1980. Generalization of selective integration procedures to anisotropic and nonlinear media. Int. J. Numer. Methods Eng. 15 (9), 1413–1418.

Kahhal, P., Brooghani, S.Y.A., Azodi, H.D., 2013. Multi-objective optimization of sheet metal forming die using genetic algorithm coupled with RSM and FEA. J. Fail. Anal. Prev. 13 (6), 771–778.

Karafillis, A.P., Boyce, M.C., 1996. Tooling and binder design for sheet metal forming processes compensating springback error. Int. J. Mach. Tools Manuf. 36 (4), 503–526.

Kaydani, H., Mohebbi, A., 2013. A comparison study of using optimization algorithms and artificial neural networks for predicting permeability. J. Pet. Sci. Eng. 112, 17–23.

Kazan, R., Firat, M., Tiryaki, A.E., 2009. Prediction of springback in wipe-bending process of sheet metal using neural network. Mater. Des. 30 (2), 418–423.

Khuri, A.I., Mukhopadhyay, S., 2010. Response surface methodology. Wiley Interdiscip. Rev. Comput. Stat. 2 (2), 128–149.

Lasdon, L.S., Waren, A.D., Jain, A., Ratner, M., 1978. Design and testing of a generalized reduced gradient code for nonlinear programming. ACM Trans. Math. Softw. (TOMS) 4 (1), 34–50.

Liao, J., Xue, X., Zhou, C., Barlat, F., Gracio, J., 2013. A springback compensation strategy and applications to bending cases. Steel Res. Int. 84 (5), 463–472.

Lingbeek, R., 2003. Aspects of a design tool for springback compensation. University of Twente/INPRO.

Lingbeek, R., Huétink, J., Ohnimus, S., Petzoldt, M., Weiher, J., 2005. The development of a finite elements based springback compensation tool for sheet metal products. J. Mater. Process. Technol. 169 (1), 115–125.

Lingbeek, R., Meinders, T., Ohnimus, S., Petzoldt, M., Weiher, J., 2006. Springback compensation: fundamental topics and practical application. In: Proceedings of Ninth ESAFORM Conference on Material Forming, pp. 403–406.

Liu, G., Lin, Z., Xu, W., Bao, Y., 2002. Variable blankholder force in U-shaped part forming for eliminating springback error. J. Mater. Process. Technol. 120 (1–3), 259–264.
Lokhande, A.M., Nandedkar, V.M., 2014. Effects of process parameters and investigation of springback using finite element analysis. Int. J. Recent Dev. Eng. Technol. 3 (1), 1–5.
Makinouchi, A., Nakamachi, E., Onate, E., Wagoner, R.H., 1993. Verification of simulation with experiment. In: NUMISHEET '93: Proceedings of the 2nd International Conference Numerical Simulation of 3-D Sheet Metal Forming Processes, verification of simulation with experiment; Isehara, Japan, August 31 to September 2, 1993.
Meinders, T., Burchitz, I., Bonte, M., Lingbeek, R., 2008. Numerical product design: springback prediction compensation and optimization. Int. J. Mach. Tools Manuf. 48 (5), 499–524.
Menezes, L., Teodosiu, C., 2000. Three-dimensional numerical simulation of the deep-drawing process using solid finite elements. J. Mater. Process. Technol. 97 (1–3), 100–106.
Montgomery, D., 2001. Design and Analysis of Experiments, fifth. John Wiley & Sons, New York, NY.
Naceur, H., Elechi, S.B., Batoz, J.L., 2005. On the design of sheet metal forming parameters for springback compensation. In: Proceedings of VIII International Conference on Computational Plasticity, pp. 1–3. Barcelona.
Naceur, H., Guo, Y.Q., Ben-Elechi, S., 2006. Response surface methodology for design of sheet forming parameters to control springback effects. Comput. Struct. 84 (26–27), 1651–1663.
Neto, D.M., Oliveira, M.C., Menezes, L.F., Alves, J.L., 2014. Applying Nagata patches to smooth discretized surfaces used in 3D frictional contact problems. Comput. Methods Appl. Mech. Eng. 271, 296–320.
Nocedal, J., Wright, S., 2006. Numerical Optimization. Springer Science & Business Media, New York, NY.
Noel, M.M., 2012. A new gradient based particle swarm optimization algorithm for accurate computation of the global minimum. Appl. Soft Comput. 12 (1), 353–359.
Oliveira, M., Alves, J., Chaparro, B., Menezes, L., 2007. Study on the influence of work hardening modeling in springback prediction. Int. J. Plast. 23 (3), 516–543.
Orme, J., Combs-Orme, T., 2009. Multiple Regression With Discrete Dependent Variables. Oxford Univ. Press, New York, NY.
Piegl, L., Tiller, W., 1997. The NURBS Book.pdf, second ed. Springer, Berlin.
Ponthot, J.P., Kleinermann, J.P., 2006. A cascade optimization methodology for automatic parameter identification and shape/process optimization in metal forming simulation. Comput. Methods Appl. Mech. Eng. 195 (41–43), 5472–5508.
Rajan, A., Malakar, T., 2015. Optimal reactive power dispatch using hybrid Nelder-Mead simplex based firefly algorithm. Int. J. Electr. Power Energy Syst. 66, 9–24.
Santos, A.D., Reis, A., Duarte, J.F., Teixeira, P., Rocha, A.B., Oliveira, M.C., Alves, J.L., Menezes, L., 2004. A benchmark for validation of numerical results in sheet metal forming. J. Mater. Process. Technol. 155–156 (1–3), 1980–1985.
Santos, A., Teixeira, P., Duarte, J., Rocha, A., Figueiredo, M., 2005. Optimizaç ao de ferramentas em estapagem: um método para a determinaç ao de relaç oes entre parâmetros do processo e a geometria obtida num componente estampado. In: Proceedings of the 4th Congresso Luso-Moçambicano de Engenharia, pp. 909–921.
Sederberg, T.W., Cardon, D.L., Finnigan, G.T., North, N.S., Zheng, J., Lyche, T., 2004. T-spline simplification and local refinement. ACM Trans. Graph. 23 (32), 276.
Serban, F.M., Bâlc, N., Achimas, G., Ciprian, C., 2013. Research concerning the springback prediction in the bending operations. Adv. Eng. Forum 8–9, 490–499.
Simpson, T., 1998. Comparison of Response Surface and Kriging Models in the Multidisciplinary Design of an Aerospike Nozzle. NASA Technical Report No. 98-16.

Steenackers, G., Devriendt, C., Guillaume, P., 2007. On the use of transmissibility measurements for finite element model updating. J. Sound Vib. 303 (3–5), 707–722.

Sun, G., Li, G., Li, Q., 2012. Variable fidelity design based surrogate and artificial bee colony algorithm for sheet metal forming process. Finite Elem. Anal. Des. 59, 76–90.

Teixeira, P., 2005. Benchmarks—Experimentais e Modelaç ao Numérica por Elementos Finitos de Processos de Conformaç ao Plástica. Master Thesis, Faculdade de Engenharia da Universidade do Porto, Portugal.

Teixeira, P., Andrade-Campos, A., Santos, A., Pires, F., César de Sá, J., 2012. Optimization strategies for springback compensation in sheet metal forming. In: Proceedings of the 1st ECCOMAS Young Investigators Conference—YIC2012, Aveiro.

Tisza, M., 2014. Advanced materials in sheet metal forming. Key Eng. Mater. 581, 137–142.

Venkataraman, P., 2009. Applied Optimization With MATLAB Programming. John Wiley & Sons, New York, NY.

Wiebenga, J., van den Boogaard, A., 2014. On the effect of numerical noise in approximate optimization of forming processes using numerical simulations. Int. J. Mater. Form. 7 (3), 317–335.

Wiebenga, J., Van Den Boogaard, A., Klaseboer, G., 2012. Sequential robust optimization of a V-bending process using numerical simulations. Struct. Multidiscip. Optim. 46 (1), 137–153.

Xu, W., Ma, C., Li, C., Feng, W., 2004. Sensitive factors in springback simulation for sheet metal forming. J. Mater. Process. Technol. 151 (1), 217–222.

Yang, X.A., Ruan, F., 2011. A die design method for springback compensation based on displacement adjustment. Int. J. Mech. Sci. 53 (5), 399–406.

Zang, H., Zhang, S., Hapeshi, K., 2010. A review of nature-inspired algorithms. J. Bionic Eng. 7 (suppl.), 232–237.

Finite element modeling of hot rolling: Steady- and unsteady-state analyses

Matruprasad Rout, Surjya K. Pal, Shiv B. Singh
Indian Institute of Technology, Kharagpur, India

4.1 Introduction

4.1.1 Flat hot rolling

Rolling is a bulk metal forming process where the material achieves the predefined thickness by plastically deforming it between two counter-rotating rolls. The desired thickness is defined by the gap between the two rolls, called the *roll gap*. From geometric point of view, though reduction in thickness of the material is the prime goal of the rolling process, different cross sections of the material can also be achieved. The shape of the product depends on the contour of the rolls. The rolling process is termed as *flat rolling* when the cross section of the material, both at input and output ends, is rectangular. *Hot rolling* is one of the basic manufacturing processes used widely to produce the flat products at working temperatures above the recrystallization temperature.

Fig. 4.1 shows a three-dimensional representation of a rolling process, and Fig. 4.2 shows its schematic representation. The presence of friction between the workpiece and the counter-rotating rolls drags the workpiece into the roll gap. As the roll gap is smaller than the input thickness of the workpiece, a product with less thickness, compared to the input thickness, is achieved. During the process, the velocity of the workpiece material, within the deformation zone, increases continuously, and at one section the velocity of the workpiece matches with the surface velocity of the roll. This section or plane is termed as *neutral plane*. The zone of deformation is divided into two zones: lagging zone and leading zone. The zone, from the entry point to the neutral plane, where the workpiece velocity is less than the roll surface velocity is termed as the *lagging zone*. Similarly, the deformation zone between the neutral plane and the exit plane, where the workpiece velocity is higher than the roll velocity is termed as the *leading zone* (Rout et al., 2015). The angle (θ) subtended by the roll, within the deformation zone, is called the *roll bite angle*. This is a critical parameter for rolling, and its value varies with the roll radius (r) and the amount of thickness reduction (h_1-h_2).

Computational Methods and Production Engineering. http://dx.doi.org/10.1016/B978-0-85709-481-0.00004-5

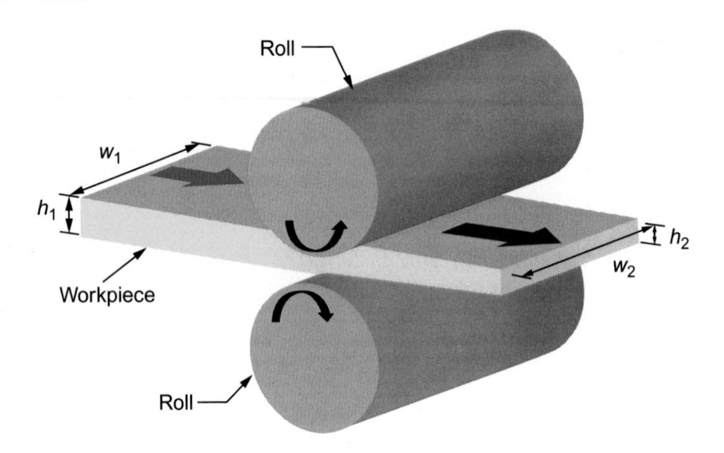

Fig. 4.1 The rolling process.

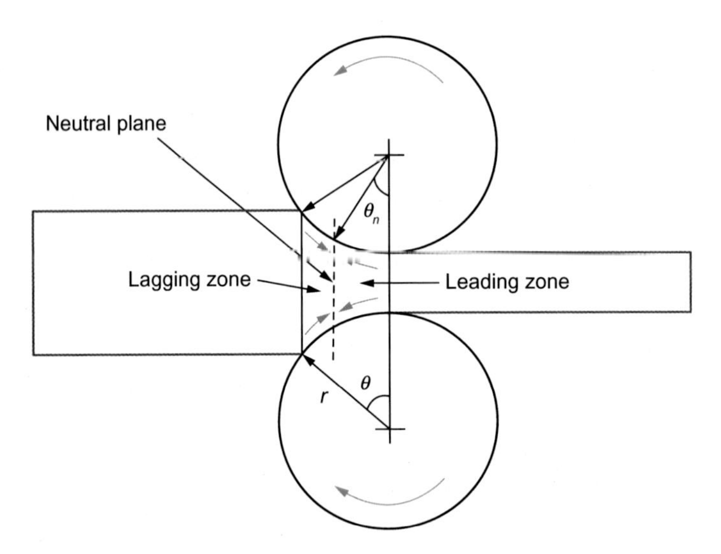

Fig. 4.2 Schematic diagram of rolling process.

4.1.2 Finite element method

The finite element method is one of the numerical tools for achieving an approximate solution to a boundary value problem, which is usually represented in the form of a partial differential equation (PDE). This method works on the principle of dividing the problem domain into smaller subdomains (called as *finite element*), representing the problem for all subdomains in the form of PDEs, assembling the PDEs of all the sub-domains, and solving the set of equations which represents the said problem for the initial domain. The term "finite element" was first introduced by Clough, around 1960, while analyzing plane stress problems within the elastic limit (Kobayashi et al., 1989). Later, it was extended to plastic deformation problems. In 1970s, finite element analysis of the metal forming processes was started (Chenot and Massoni, 2006). The finite element method is also known as finite element modeling (FEM) or finite element analysis (FEA).

4.1.3 FEA of hot rolling process

Development of fast computing devices has made the application area of FEM quite broad. Also, in terms of cost, it has advantages over experimentation. FEM has proved to be a powerful, cost effective, and reliable tool for analyzing various manufacturing processes. A single model in FEM can predict many variables such as velocity, stress, strain, temperature, etc. Continuous evolution of manufacturing processes to improve the product quality and to reduce the cost of manufacturing leads to the evolution of FEM programs in the form of commercial computer codes. ANSYS, ABAQUAS, LS-DYNA, DEFORM-3D, and QFORM are some of the commercial software used for FEA of metal forming processes. Prior to FEM, analytical models were used to solve various problems in rolling. In general, a mathematical model of the flat rolling process includes:

- equations of motion of the deformed metal
- heat balance of the roll/strip
- equations of equilibrium of the work-roll
- description of the interobject relationship between the work-roll and the deformed metal
- description of the material properties

These mathematical models are used to predict various state variables such as roll pressure distribution, roll separating force, roll torque, etc. Mathematical models, written in a programming language, are used for FEA of various processes. Application of FEM in rolling process can vary in many respects (Galantucci and Tricarico, 1999; Shahani et al., 2009):

- type of approach: transient or steady state
- formulation type: solid formulation or flow formulation
- technique for solution: Lagrangian, Eulerian, arbitrary Lagrangian-Eulerian (ALE)
- material behavioral constitutive law: elasto-plastic, elasto-viscoplastic, rigid-plastic, rigid-viscoplastic
- type of discretization: two-dimensional or three-dimensional
- analysis type: thermal, mechanical, uncoupled, or coupled thermomechanical

In addition to thermomechanical study, various metallurgical features can also be incorporated to FEM for study of grain evolution, texture development, etc. While doing FEA for hot rolling, lot of assumptions are considered that may not be valid in real industrial rolling. So, realistic assumptions must be made. Most of the researchers validate their simulation results with those obtained from industry, while some validate against experimental results obtained on a laboratory scale. Though this has got limitations such as low roll speed, smaller sample size, etc., these models are also applicable to large-scale industrial rolling.

Discretizing the domain to small elements, i.e., meshing of the domain is the first step in FEA that has to be done after the model is created. Based on this mesh, the rolling process can be analyzed with a fixed mesh, i.e., Eulerian approach or with a mesh which needs to be updated with deformation, i.e., Lagrangian approach. Usually, Eulerian approach is appropriate for steady-state analysis and Lagrangian approach is for non-steady- or transient-state analysis. In metal forming, where bulk deformation takes place, the elastic deformation can be neglected so that rigid-plastic material model can be used. But, in situations where stress rate is required, like for prediction of thermal stresses, the elastic deformation has to be included in the model. Thus an elastic-plastic model needs to be considered. Similarly for materials sensitive to deformation rate or strain rate, either rigid-viscoplastic or elasto-viscoplastic model is selected. FE modeling of hot forming problems can be carried out by coupled or uncoupled approach. In the uncoupled analysis, the result of thermal model becomes the boundary conditions for the mechanical or deformation model, whereas in coupled analysis, the temperature and the deformation, i.e., the velocity fields are solved together. In flat rolling, the width of the sample is quite large in comparison to the thickness due to which rolling is normally considered as a plane strain process (Shangwu et al., 1999). For these situations, two-dimensional analysis can give good results and it also reduces the computational time. But for cases where there is a considerable spread in the width direction, like bar or rod rolling, the assumption does not hold good and a three-dimensional analysis is required in order to get the overall picture of the deformation. The next section describes the research work carried out in the field FEA of hot rolling process to study the thermomechanical behavior of the workpiece.

4.2 Thermomechanical analysis of hot rolling: An overview

The early research works on FEA of rolling were carried out by Tamano (1973), Zienkiewicz et al. (1978), Dawson (1978), Kanazawa and Marcal (1978), and Shima et al. (1980). Tamano used an elastic-plastic material model, whereas Zienkiewicz et al. (1978) and Dawson (1978) used viscoplastic material model. Shima et al. (1980) used the rigid-plastic model to study different cases of rolling, but their analysis did not consider the work-hardening behavior of the material. All these analyses were made considering the plane strain situation. The spreading along the width direction was first considered by Kanazawa and Marcal (1978), whereas the heat generation due to plastic deformation was incorporated by Zienkiewicz et al. (1978). In 1980s Kobayashi et al. (Kobayashi et al., 1989; Kobayashi, 1982; Li and Kobayashi, 1982) made a significant contribution to the field of analytical and finite element analyses of metal forming processes. Li and Kobayashi (1982) investigated the plane strain rolling, for a wide variety of materials, by using rigid-plastic FEM based on variational approach. They studied both the steady and unsteady cases of rolling. Velocity-dependent friction boundary was considered. They have suggested a parameter, H, to evaluate the effect of reduction per pass and roll gap geometry on deformational homogeneity. The existence of single and double peaks in the pressure distribution curve was also correlated to the parameter H. Kobayashi (1982) summarized the application of two-dimensional FEM in metal forming. He concluded that along with the process analysis, friction condition, residual stress distribution, constitutive model, plastic instability, and relation between deformation mechanism and microstructure development are the other critical areas in the field of hot forming. The earlier mentioned work along with some other research done by Thompson (1982), Mori and Osakada (1984), Liu et al. (1987), Lau et al. (1989), Lindgren and Edberg (1990), and Kim and Kobayashi (1990) was carried out for cold rolling. The basic steps of carrying out FEA for hot rolling is the same as that for cold rolling, but the main factor relevant to hot rolling is the constitutive equation, i.e., the material behavior at higher temperature. Also, the friction factor and the heat transfer between the roll and the workpiece are two other important factors that significantly affect the simulation results. So, to carry out FE simulation of hot rolling process, adequate knowledge about material model, friction condition, and heat transfer at the roll-workpiece interface should be known. Thus application of FEA to hot rolling includes the following areas:

- thermomechanical behavior of the rolled plate during rolling
- microstructural evolution during rolling
- work-roll and plate interface behavior (friction and lubrication)
- crack formation in the rolled plate during rolling
- thermomechanical behavior of the work-roll
- thermal stress analysis of the rolled plate on run-out table

The present chapter focuses on the deformation and thermal analyses of the plate in the deformation zone and hence, literature related to these aspects is reported here.

A simple three-dimensional thermomechanical model, based on flow formulation, was used by Montmitonnet et al. (1992) to study the hot rolling process. The material behavior was expressed by Norton-Hoff viscoplastic model and the predicted stress-strain field in the billet was used to study the evolution of surface cracks. The authors also studied the spread in the billet. Effect of anisotropic behavior of material properties in FEA of hot rolling had been studied by Montmitonnet et al. (1996). They have considered a three-dimensional model to study the effect of anisotropy on stress distribution and spread in the metal and found that both are quite sensitive to anisotropy. They also reported that by considering the anisotropy formulation, the CPU time increases by 30%–50%. Dvorkin et al. (1997) investigated the transient- and steady-state plane strain hot rolling process by considering a two-dimensional model, based on flow formulation. The model demonstrates the slipping situation and downward bending of the rolled plate. The slipping situation was analyzed by the steady-state model by varying the friction coefficient and the study was done for both the cases, with and without tension between two stands. The downward bending of the rolled plate due to the asymmetric plate temperature was analyzed by the transient model. Lin and Shen (1997) proposed a two-dimensional finite element model which simultaneously solves the deformation and heat transfer problem of the element. The obtained rolling force and temperature distribution were compared with the previously reported experimental results. A two-dimensional coupled thermomechanical FE model has been proposed by Galantucci and Tricarico (1999) to analyze the roughing and finishing rolling processes. They expressed the rolling parameters in a parametric form, which gives flexibility to the model. The transient plane strain rolling process was analyzed by considering the heat generated due to friction, as well as plastic deformation.

In the beginning of the 21st century, there is a significant development in the research field of FEA of hot rolling process, and quite a good number of research works have been published. Development of high performance computers provides the platform to handle large and complicated problems and also reduces the computational time. In the meantime, development in the field of commercial FEA software also contributed significantly, like analysis of large deformation became possible due to the ability of the software to handle the mesh distortion effectively. Cavaliere et al. (2001) studied the rolling process for steel, where they coupled the deformation model of the workpiece with the thermomechanical model of work-roll. The plate was described as an Eulerian rigid-viscoplastic model whereas the work-roll was defined as a Lagrangian elastic model. With this coupled model, they have studied the deformation of roll, its temperature distribution and thermal expansion. The effect of work-roll crown on plate profile and roll separating force was also reported. In hot rolling, as the hot plate comes in to contact with the relatively cool roll, the temperature of the plate drops drastically, which is known as roll chilling effect. This produces an inhomogeneous temperature distribution in the through-thickness direction of the plate. However, this effect is limited to a layer of 1/15th to 1/20th of the thickness, as reported by Serajzadeh et al. (2002). Friction between work-roll and plate is a desirable parameter for rolling to occur, but at the same time, a high value in friction also increases the rolling load. Tieu et al. (2002) considered a three-dimensional rigid-viscoplastic finite element model to study the hot rolling process with lubrication. They conducted hot rolling experiments with and without lubrication. The friction factor was calculated by using an empirical formula based on measured distribution of roll pressure and used the same in FEM to study the effect of lubrication. Simulation for the case of lubrication results in decrease in roll separating force and roll torque. Synka and Kainz (2003) developed a new mathematical model for the simulation of steady-state hot rolling process. They defined the deformable body by a mixed Eulerian and Lagrangian coordinates. In their model, Eulerian coordinate is used to define the body in rolling direction whereas width and thickness directions were defined by Lagrangian coordinate. Solutions obtained for displacement, velocity, and stress field from this model were in good agreement with other mathematical models. Byon et al. (2004) considered a three-dimensional Eulerian FE model to study the deformation of the plate in a steady-state hot rolling process. The constitutive model for the deformable rigid-viscoplastic material was developed considering the dynamic softening of the material by dynamic recrystallization (DRX) and dynamic recovery (DRV). They concluded that softening of the material within the roll gap plays an important role in predicting the rolling force accurately. Sun (2005) implemented FE to study the behavior of the roll-workpiece interface. Both the parameter, i.e., the friction coefficient and interfacial heat transfer coefficient were described as a function of roll pressure and mean flow stress. They have studied the effect of various process parameters such as workpiece temperature, roll speed and amount of reduction on rolling force, pressure distribution and hence, on friction coefficient and interfacial heat transfer coefficient. The variations in friction factor and interfacial heat transfer coefficient were also explained by the effect of scale formation, effect of temperature, and viscosity of lubricant.

Measurement of the temperature inhomogeneity inside the material in the deformation zone is a tough task as rolling is a dynamic process and also involves large plastic deformation. It is not feasible to measure the temperature at the top surface of the workpiece in the deformation zone, and hence noncontact mode of measurement of temperature is preferred. But, this mode of measurement will not give the temperature of the workpiece immediately below the work-roll. Riahifar and Serajzadeh (2007) carried out rolling experiment with K-type thermocouple to validate their FEA result for hot rolling of aluminum alloys. The thermocouples were embedded in the sample at the mid-thickness and at the subsurface. Exact location of thermocouple was not reported. They have also studied the lateral spread and through-thickness stress-strain distribution of the workpiece within the deformation zone. Inhomogeneity in temperature and strain distribution leads to a nonuniform effective strain distribution along the thickness direction. The reported result shows positive principle stress distribution at the edge of the workpiece, making these areas prone to surface crack. Pal et al. (2007) approached the hot rolling problem by Bond graph method. The authors successfully implemented the method in predicting temperature distribution in the plate during hot rolling and also got a close match with the FEM result. They concluded that a well-developed bond graph approach can be a useful tool in analyzing the rolling process. Nonlinear rigid-viscoplastic FEM has been used by Min ting et al. (2007) to simulate the hot strip continuous rolling process. They have considered softening by recrystallization processes such as DRX, MRX (metadynamic recrystallization), and SRX (static recrystallization) to predict the final grain size of the rolled product. The model gave good results for grain size and other outputs like rolling force and temperature. Esteban et al. (2007) studied the effect of friction coefficient and looper stress on lateral spread and power consumption during a rolling process. The looper stress is the stress developed by the looper mechanism used to keep the slab in tension between two stands. They first formulated the constitutive equations for the workpiece material by torsion test at different temperatures and strain rates. The developed material model was then used to study the rolling process by FE modeling. The simulations were carried out by neglecting the thermal losses. With increase in the coefficient of friction, the difference in the spread value between the central plane and the top surface was found to increase. Similarly, power consumption also increased with the increase in the coefficient of friction. The effect of looper stress was less significant in both the cases. Serajzadeh and Mahmoodkhani (2008) proposed a two-dimensional mathematical model based on upper bound theorem and FEA. The velocity field, predicted by the upper bound method and coupled with the two-dimensional FE model, was used for thermal analysis. Predicted temperature distribution was validated with the temperature, measured by thermocouples embedded in the workpiece. The flow pattern of the material, obtained from FEA, was compared with the flow pattern of experimentally rolled workpiece. They constructed rectangular grids on the side edges of the workpiece to study the deformations. They found same pattern of deformation for both the distorted FE mesh and the rectangular grids on the experimentally rolled workpiece. Effect of various rolling parameters was also reported. The authors also concluded that by using a combined upper bound and FE methods, the simulation time is reduced as compared with a single FE model. Sheikh (2009) also studied the thermal behavior of the strip in hot rolling process by using upper bound technique and FEA. Yuan et al. (2009) analyzed the thermomechanical behavior of the billet for 18-pass continuous hot rolling process. They have implemented a pushing technique of a rigid body, which pushes the deformable billet into the next stand. The velocity of the rigid body was defined by the front and tail end velocities. During deformation, the velocity of the rigid body was same as that of the billet tail end, whereas after the deformation of the billet, the rigid body gets the velocity of the billet front end, which helps in moving the billet to the next stand. To tackle the mesh distortion problem in large deformation, they have used different models for different stands. The new model, for the next stand, has the same shape as that of the previously deformed model, but with a new mesh. Data mapping technique was used to map the data from the deformed mesh to the new mesh. This helps in making the input of the new model same as that of the deformed model. By implementing these two techniques, they reported well predicted thermal response of the billet. They also reported that the pushing technique reduces the overall computation time for the multipass rolling process.

Though FEA is a reliable method to analyze hot rolling process, it may not be useful for online application because of its high computation time. As a solution for this, some researchers implemented artificial neural network (ANN) to the hot rolling process. Shahani et al. (2009) used an ANN technique to predict the behavior of the slab during hot rolling process. They first carried out thermomechanical analysis of the slab for two-dimensional steady-state hot rolling process by FEA and compared the results with the previously published research work. Output of the FEA was used to train the ANN model which later predicted the effect of various process parameters for the hot rolling process. The implemented ANN model was designed based on back-propagation technique. The computation time, taken by FEA, depends on the convergence speed, which depends on the initial guess for the solution of the problem under consideration. Zhang et al. (2009) considered three different methods, viz., engineering method, G functional method, and neural network (NN) to predict the initial guess of the two-dimensional rigid-plastic FEM for seven-pass hot rolling process. The initial guess values for velocity and energy, obtained from different methods, were compared with the real velocity and energy achieved by the FEM. The results were also compared in terms of number of iterations and time taken by the CPU. Based on the comparisons, the authors concluded that rigid-plastic FEM gives better results for the initial guess predicted by NN. A similar kind of research work was reported later by Zhang et al. (2010). They studied the rigid-plastic FEM for online application in strip rolling, where the initial guess was predicted by NN. The FE model, with less than 2500 elements, took 500 ms and predicted rolling force value with an accuracy of $\pm 10\%$.

Bagheripoor and Bisadi (2011) used ALE formulation to study the effect of different rolling parameters on the temperature distribution of AA1100 alloy plate. They considered the material as a thermo-viscoplastic and the material behavior was described by hyperbolic sine equation. Among the rolling parameters, they considered roll speed, amount of reduction, initial thickness of the plate, and heat transfer coefficient. The authors used plant recorded data to verify the predicted results. Ben et al. (2011) simulated the hot rolling process for steel plate by neglecting the heat conduction in width and rolling directions. For rolling load and plate temperature, they obtained a good agreement with the plant recorded data.

The earlier mentioned overview shows that the major process parameters which significantly affect the FEM predicted results are the work-roll-workpiece interface behavior and the constitutive equation which defines the material behavior. So, a brief introduction to these parameters is given as follows.

4.3 Work-roll and workpiece interface behavior

The interobject relation, i.e., the contact boundary condition between work-roll and workpiece is specified by the interfacial heat transfer coefficient (IHTC) and type of friction. Both of them are critical from numerical analysis point of view to get an accurate result. IHTC is more critical for hot rolling as it directly controls the workpiece temperature and hence the required rolling load and evolution of microstructure. Measurement of both of these parameters is very difficult as the situation is dynamic and the time of contact between work-roll and workpiece is very small (of the order of some fraction of second). Heat transfer analysis in the roll gap is therefore a critical area of research for the past few decades (Chen et al., 1992, 1993).

Deformation of the material is a very complicated process. The formation of oxide layer on the workpiece at high temperature further increases its complexity. Polozine and Schaeffer (2008) summarized the different factors which affect the heat transfer in the contact region for forging and the same factors are also valid for hot rolling. The factors are:

- presence of oxide film and its thickness in the contact zone
- layer thickness of lubricant between the work-roll and workpiece
- amount of air trapped in the contact zone
- roughness of the work-roll and workpiece surfaces
- temperature of work-roll and workpiece
- normal pressure at the contact
- thermophysical properties of the lubricant, oxide film, work-roll, and workpiece materials

Moreover, the thermal analysis of the workpiece in the deformation zone can be summarized as follows:

- heat transfer from the workpiece to the environment by radiation and convection
- heat generation because of plastic deformation of the material
- conductive heat transfer from the workpiece to the comparatively cool work-roll
- heat generation because of the friction between work-roll and workpiece

Heat transfer from the workpiece to the environment can be specified by the fundamental heat transfer law of radiation and convection, as follows (Pal et al., 2007):

$$q_t = h_\infty (T_s - T_\infty) + \varepsilon\sigma (T_s^4 - T_\infty^4) \tag{4.1}$$

where, h_∞ is the convective heat transfer coefficient, T_s is the surface temperature of the workpiece, T_∞ is the temperature of the environment, ε is the emissivity of the workpiece material, and σ is the Stefan-Boltzmann constant.

Significant amount of heat is generated when metal is deformed plastically, and its amount increases with the increase in deformation rate as the time available for heat to conduct to other zone of material is less. The increase in temperature softens the material and hence influences the mechanical properties of the metal. A very small fraction of this plastic work is stored in the material and the rest is converted to heat energy. The conversion of mechanical to heat energy can be found out by using the equation, $q_m = \hat{n}\overline{\sigma}\dot{\overline{\varepsilon}}$ where $\hat{n}$ is the fraction of plastic work converted to heat and is usually taken as 0.9 (Kiuchi et al., 2000), $\overline{\sigma}$ is the effective stress, and $\dot{\overline{\varepsilon}}$ is the effective strain rate.

4.3.1 *Interfacial heat transfer coefficient*

The temperature drop in the workpiece as and when it touches the relatively cold rolls takes place only in a small subsurface volume and can be regained by the conduction of heat from the central part of the workpiece. The heat transfer from the workpiece to the work-roll is through conduction, but identifying a proper value of IHTC is very difficult. Considering different factors affecting the heat transfer in the deformation zone and complexity of the deformation, the Fourier law, which describes the heat transfer by conduction in a homogeneous medium, cannot be applied here (Polozine and Schaeffer, 2008). Most of the researchers have defined it in a simplified manner and tried to estimate the heat transfer coefficient by comparing the experimentally measured rolling load and temperature values of the plate with the numerical analysis. Literature on heat transfer in roll gap shows that the IHTC is not constant throughout the deformation zone and increases from entry to exit zone (Chen et al., 1993). It also shows that the value of IHTC increases with the increase in the contact normal pressure (Devadas et al., 1991). Formation of oxide layer on the workpiece also has a significant effect on the rate of heat transfer to the work-roll. The IHTC value usually decreases with increase in oxide layer thickness (Hu et al., 2013). However under same condition (workpiece thickness, reduction, speed, and deformation temperature), the IHTC value in hot rolling is higher than that of hot forging (Li and Sellars, 1998). Literature shows that a wide range of IHTC values, from 2 to 200 kW/m^2 K, is considered for the numerical analysis for hot rolling of various kinds of steel (Sellars, 1985). However, for stainless steel, a higher IHTC (up to 620 kW/m^2 K) value has also been reported (Chen et al., 1993).

4.3.2 Friction coefficient

Two objects, in contact, always experience some amount of friction at the contact surface. In order to move an object relative to the other one, the resistance offered at the interface must be overcome. Similarly, for the case of rolling the following condition must be satisfied for rolling (Fig. 4.3):

$$F\cos\theta \geq P\sin\theta \tag{4.2a}$$

$$\mu \geq \tan\theta \tag{4.2b}$$

where F is the frictional force, P is the normal pressure, μ is the coefficient of friction, and θ is the angle of bite. Fig. 4.3 shows the components of force acting at the contact point between the roll and the workpiece.

The importance of friction in the rolling process cannot be overlooked. In the lagging zone, i.e., when the workpiece enters the roll gap, the surface speed of the rolls is higher than that of the workpiece. So, the direction of friction is in the direction of the workpiece movement and this draws the workpiece into the roll gap. The velocity of the workpiece increases as it moves and at one plane in the roll gap (called the neutral plane) it becomes equal to the surface velocity of the rolls. Beyond this plane, i.e., in the leading zone, the billet speed is faster than the peripheral speed of the rolls and direction of friction reverses. The coefficient of friction at the interface can be expressed as the ratio of frictional force (F) and normal pressure (P), i.e., Coulomb friction or by the ratio of interfacial shear stress (τ) and shear yield strength of the material (k_1), i.e., shear friction. Though friction helps in rolling, a higher value of friction

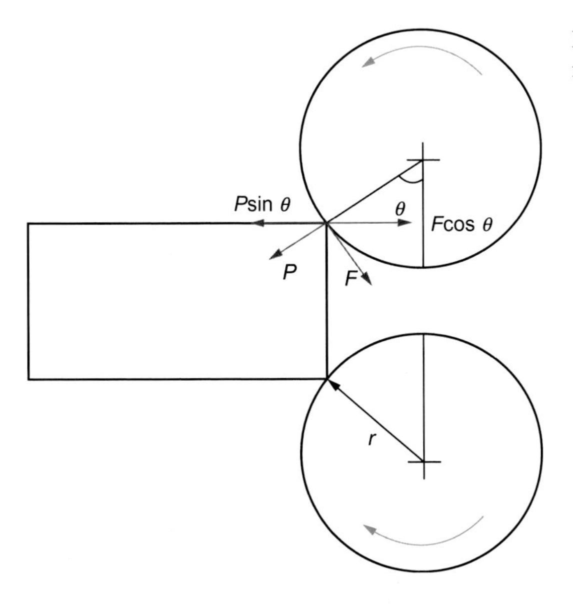

Fig. 4.3 Frictional force during rolling.

factor increases the rolling load. Frictional force at the interface also affects the surface quality of the rolled produced. Hence, lubrication at the interface is necessary.

The importance of friction in the rolling process motivates researchers to analyze the deformation zone for various lubrication conditions. Analysis of deformation zone is based on some assumptions. Orowan (1943) has summarized the assumptions made in earlier theories of cold rolling process as follows:

- during rolling, planes perpendicular to the rolling direction remain in plane
- coefficient of friction is constant over the contact area
- no lateral spreading of the rolled material
- rolled material obeys von Mises' yield criterion
- contact arc is assumed as circular and elastic deformation of the rolled material is negligible

Experimental methods (Theocaris, 1985) are also available for finding out the friction factor. However, in hot rolling, the interface behavior is much more complicated and the friction factor is higher as compared to cold rolling process. The materials under deformation are strain rate dependent and the lateral spread is also substantial. Moreover, the working environment is at high temperature. So, accurate measurement of coefficient of friction is difficult, which leads to the assumption of a constant coefficient of friction throughout the deformation zone. However, the effect of various parameters on coefficient of friction cannot be neglected. Though a constant coefficient of friction is considered, it must reflect all those effects. Various parameters that affect the friction coefficient are:

- surface roughness of work-roll and workpiece
- temperature of the workpiece
- type of lubricant and its thickness
- thickness of oxide layer

4.4 Constitutive equation for material model

Constitutive equation or model is a mathematical formulation which establishes a quantitative relation between two parameters and hence characterizes the material properties. It describes how the material responds to external stimuli. The parameters may be stress-strain, temperature-heat flux, current-voltage, etc., depending upon the problem definition. In the present case, it is the stress-strain relationship or the flow stress model. As mentioned earlier, at higher temperature a variety of metallurgical phenomena may take place and the material may become more sensitive to strain rate; thus the constitutive equation becomes more complex. At the same time, formulating a single model that is accounting for all the effects of strain, strain rate, temperature, and other metallurgical phenomena on the flow stress is also difficult. Out of all the available models till date, the constitutive model which best describes the material behavior at higher temperature and over a wide range of strain rates was introduced in the 1960s (McQueen and Ryan, 2002) and is represented in Eq. (4.3).

$$Z = \dot{\bar{\varepsilon}} \exp\frac{Q}{RT} = A[\sinh(\alpha\bar{\sigma})]^n \tag{4.3}$$

Table 4.1 **Constitutive model used by different authors**

Constitutive model	Applied by authors
Based on Zener-Hollomon parameter	Rout et al. (2016a), Ding et al. (2015), Sui et al. (2013), Ding et al. (2013), Bagheripoor and Bisadi (2011), Huang et al. (2011), Ding et al. (2010), Serajzadeh and Mahmoodkhani (2008), Esteban et al. (2007), Pal et al. (2007), Riahifar and Serajzadeh (2007), Byon et al. (2004), and Aiyedun et al. (1997).
Norton-Hoff law	Riahifar and Serajzadeh (2007), Montmitonnet (2006), Duan and Sheppard (2004), Duan and Sheppard (2002), Montmitonnet et al. (1996), and Montmitonnet et al. (1992)
Hensel and Spittel equation	Brand et al. (1996), Synka and Kainz (2003), Duan and Sheppard (2004), and Nalawade et al. (2013)
Shida's model[a]	Byon et al. (2013), Phaniraj et al. (2005), Yang et al. (2003), Lee (2002), Sun and Hwang (2000), Krzyzanowski et al. (2000), and Said et al. (1999)
Misaka's model[a]	Byon et al. (2013)
Perzyna equation	Shahani et al. (2009)
Model with linear hardening law	Vallellano et al. (2008)

[a]The Shida's model and Misaka's model contain the percentage of carbon as a variable, and thus can be used for different grades of steel.

where A, n, and α are material constants; Z is the Zener-Hollomon parameter, Q is the activation energy, R is the gas constant, and T is the absolute temperature. Constitutive model containing Zener-Hollomon parameter describes the flow stress most effectively and hence has been used by many researchers. Table 4.1 presents the constitutive models used by different authors.

4.5 Basic steps of FEM

Most of the problems faced in engineering can be modeled mathematically, and hence, can be solved by employing various available mathematical techniques. Engineering problems are basically the boundary value problems which can be described by one or set of differential equations and/or algebraic equations. These equations may contain one or more dependent and independent variables. The physical structure, in real scenario, becomes the field of interest or the working domain and the existing problem in terms of the differential equation should satisfy throughout this domain. The dependent variables, termed as field variables, have specified values at the boundaries and are referred to as boundary conditions. Solution of such engineering problems by FEM involves the following basic steps:

- discretization of the domain
- defining the element by shape function

- deriving the element stiffness equation
- formation of global stiffness matrix
- incorporating boundary conditions
- solving the equations

Details of the earlier mentioned steps can be found from any FEA textbook (Reddy, 2012; Hutton, 2004). However, a very brief description to the earlier mentioned steps is given here.

Obtaining an approximate solution to the problem, through FEM, starts with discretizing or dividing the working domain into finite number of smaller units. The process of forming smaller units is called *meshing* and each small unit is called an "*element.*" Higher the number of elements (lesser the element size) better is the accuracy. However, the computational time increases with increase in the number of the elements. Also the number of elements used to discretize a domain depends on the type of elements. So, an optimized value for the element size should be chosen.

Discretizing the working domain into subdomains helps in converting the complex geometry into comparatively simple geometry and also increases the accuracy of the approximated solution as compared with that of a single domain. The types of elements used generally are:

i. one-dimensional analysis: bar element
ii. two-dimensional analysis: triangular and rectangular elements
iii. three-dimensional analysis: tetrahedron and brick elements

The elements are interconnected to each other at *nodes*. The defined problems in the form of an equation are solved at these node points. However, the equation needs to be satisfied over the whole domain and for the same, *interpolation function* or *shape function* is used. This shape function is described in the form of polynomials containing the field variables and is defined within the element so that the continuity of the solution throughout the domain can be maintained. The governing equation which needs to be solved within the element first and then over the whole domain is defined by a *weak form*. A weak form is a weighted integral statement of governing differential equations. In this form, from dependent variables, the differentiation is transferred into integral weight function such that all natural boundary conditions are included in the integral. Methods like variational and weighted residual are normally used to get the weak form (integral form) from the strong form (differential form) governing equation. Obtaining weak form through variational approach is based on either principle of minimum potential energy, principle of virtual work or Castigliano's theorem, whereas least square method, Galerkin's method, Petrov-Galerkin method, and collocation method are the well-known methods used for the derivation of weak form by weighted residual method. Once weak form is obtained, the shape function is put into this weak form integral equation to generate the required equations of finite elements, which need to be solved in further steps. This is to be noted that along with the governing equations, in terms of field variables, constitutive equations for the material model need also to be solved (Priyadarshini et al., 2012; Jain et al., 2015).

Method of superposition is used to combine the individual element equation to form the *global stiffness matrix*. The boundary conditions are then incorporated to get the final set of equations. These equations are then solved through iterative methods, e.g., direct iteration method, Newton-Raphson method, etc.

4.6 Different approaches in FEM

Metal forming problems are highly nonlinear and also involve large plastic deformation. Large deformation may lead to mesh distortion and hence, FEM becomes computationally inefficient. So, selection of proper formulation for the problem plays an important role to get an efficient result. Problems in FEM involving motion of deformable body can be formulated by three different approaches (Priyadarshini et al., 2012):

- Lagrangian approach
- Eulerian approach
- ALE approach

These approaches are based on the behavior of the mesh system used, i.e., the computational reference mesh system on which the calculations are made and the material reference mesh system, which follows the deformed material (Li et al., 2009). A brief introduction to the earlier mentioned approaches has been given in the following subsections.

4.6.1 Lagrangian approach

In this approach, the nodes are attached to the deformable body and move as per change of shape caused by deformation. Nevertheless, the relative position of the nodes with respect to the material points always remains fixed. This approach treats the material with history-dependent constitutive relations. As the nodes are attached to the material, variation of state variables such as temperature, velocity, stress, strain, etc., throughout the process, can be easily predicted. Also implementation of the boundary conditions is simple. This approach suffers from two main drawbacks: mesh distortion and high computational time. For large deformation problems, deformation of mesh with the material may lead to mesh distortion. This makes this approach limited to structural and comparatively smaller deformation problems. However, with a proper remeshing criterion, this approach can also be used for large deformation problems. Similarly, high computational time can be tackled by proper selection of solver and iteration method. Also the computational time can be largely reduced by implementing parallel computation (Chenot and Massoni, 2006).

The Lagrangian approach to solve the problem can be adopted in two different ways: *total Lagrangian* (TL) and *updated Lagrangian* (UL). For the first type, the weak form involves integrals over the initial configuration, whereas for the second type the integrals are over the deformed configuration. In other words, for TL approach, all the variables correspond to the initial configuration, and for UL approach, they correspond to the last deformed configuration, and hence the name updated Lagrangian. However, the derivatives in TL are with respect to the material coordinates, but in UL these are with respect to the spatial coordinates. Both the approaches represent the same mechanical behavior and the choice of formulation is a matter of convenience (Priyadarshini et al., 2012).

4.6.2 Eulerian approach

In the Eulerian approach of solving FE problems, the nodes corresponding to the working volume need to be analyzed and are fixed in space. In other words, a control volume is defined and the flow of material through this volume is analyzed. As the nodes are fixed, problems related to mesh distortion are removed. But at the same time the nodes may not coincide the element edges, leading to difficulties in defining the boundary conditions (Priyadarshini et al., 2012; Jain et al., 2015). This also increases the complexity in handling the material constitutive equations. However, the Eulerian approach has the advantage of shorter computational time as compared with Lagrangian approach. For hot rolling, to study the thermomechanical processing of the work material, deformation zone in the roll gap is the control volume.

4.6.3 Arbitrary Lagrangian-Eulerian (ALE) approach

The ALE approach, as its name indicates, is an attempt to achieve the advantage of both the Lagrangian and the Eulerian approaches. In this approach, the nodes in the mesh can be flexible and move as the material deforms like the Lagrangian mesh or can be fixed in space and eliminate the mesh distortion issues like Eulerian mesh or it can move in some arbitrary manner and give a continuous rejoining capability. This flexibility of the mesh, offered by ALE description, makes it capable of handling large distortion problems.

To study the in-process workpiece geometry evolution or material flow during the process, perhaps, the Lagrangian approach is the best one. But, its high computational time and inability to handle the mesh system during high deformation make Eulerian approach more suitable. Also sometime it is necessary to do a few preliminary simulations before going for simulation of the actual process. In those cases Eulerian approach is more appropriate because of its less computational time.

4.7 Solution methods

The metal forming problems are nonlinear and dynamic in nature. Also the processes are transient. So, the numerical analysis of these problems involves transient nonlinear dynamic set of equations which can be analyzed implicitly or explicitly. In implicit analysis, the variables are solved by solving equations involving the current state (at time, $t_{n+1} = t_n + \Delta t$) and the previous state (at time, $t = t_n$) of the system. This type of time integration makes the solution unconditionally stable, and hence a larger time step can be considered. Also this method needs inversion of the stiffness matrix, which is time consuming. Larger the stiffness matrix higher will be the computational time.

Table 4.2 **Roll mill details (at IIT Kharagpur)**

Type	Capacity	Roll diameter	Roll width	Rotational speed
2 high reverse rolling mill	75 ton	320 mm	300 mm	18 mpm (max.)

Table 4.3 **Sample details**

Material	Avg. dimension ($W \times H \times L$) mm
304 austenitic stainless steel	$11 \times 20.4 \times 78$

So, this method is usually preferred for problems involving linear equations, because for nonlinear problems the degree of freedom is higher with larger stiffness matrix. However, this method can be efficiently implemented to solve nonlinear problems by converting them to linear equations. On the other hand, in explicit analysis the unknown variables for the current state (at time, $t_{n+1} = t_n + \Delta t$) are calculated from the values of previous step (at time, $t = t_n$). This analysis does not require the inversion of stiffness matrix which significantly reduces the computational time, but the time step required for this analysis is very small to keep the solution stable. So, the choice of solution method depends on the kind of problem to be solved.

In subsequent sections focus is on FEM of the hot rolling process by both the approaches: steady state and unsteady state. The roll dimension has been considered based on the available laboratory scale rolling mill at Indian Institute of Technology (IIT) Kharagpur, India. Comparatively smaller sample dimensions have been taken in order to reduce the computational time as the process is analyzed as a three-dimensional problem. However, for industry-scale hot rolling or cases where plane strain situation is valid, two-dimensional analysis will be more appropriate and efficient as well. DEFORM-3D platform, which solves the finite element problems with an implicit solver, has been considered to carry out the present work (Tables 4.2 and 4.3).

4.8 Steady- and unsteady-state analyses of hot rolling

The hot rolling process is usually a multipass process, where a number of roll stands (set of two rolls) with continuously decreasing roll gap is arranged in line. For each stand, except at the beginning and the end of the deformations, the rolling process is a kinematically steady-state process (Li and Kobayashi, 1982). Here, kinematically steady state means the process configuration does not change with respect to time. So, the deformation as well as the heat transfer in a hot rolling process can be analyzed as a steady-state problem. Since the process configuration remains same with time, a fixed mesh is suitable. But, at the same time the solution to the earlier mentioned analysis is dependent on loading history. So, the considered mesh should have the ability to update continuously according to the external loading. In other words, the steady-state analysis of hot rolling process needs an ALE approach; whereas, for the unsteady- or transient-state analysis, a movable mesh with continuously updating ability is required. Hence, an updated Lagrangian approach is more appropriate (Kobayashi et al., 1989). Apart from the formulation, some differences also exist in the initial geometry consideration, the initial mesh, and the boundary conditions. These are discussed in detail in the following subsections.

4.8.1 Initial geometry

4.8.1.1 Steady-state analysis

In this approach, a part of the sample that is under the steady-state condition during rolling is considered. For this, the initial geometry of the sample is modeled with the deformation zone geometry, which depends on the geometry of the roll. So, the initial sample has both inlet and outlet thicknesses along with the varying thickness in the deformation zone (Fig. 4.4A).

4.8.1.2 Unsteady-state analysis

In unsteady state or transient analysis of hot rolling, the actual rolling process is analyzed and hence, the initial sample to be modeled has the dimension same as that of the input material (Fig. 4.5A). The process is simulated starting from the sample entering to the roll gap to the exit of the sample from the roll gap.

This is to be noted that the considered sample geometry for the present study has a symmetry along both the axes: X and Y. So, half or quarter of the sample geometry can be considered for the analysis to reduce the computational time. However, to get the

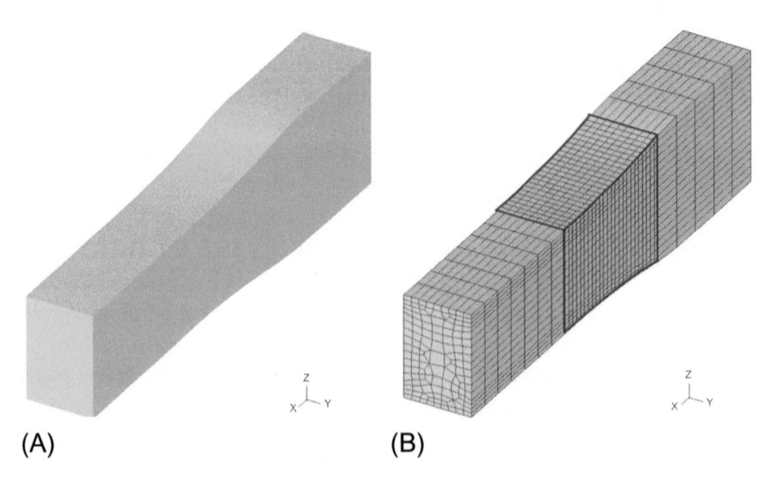

Fig. 4.4 Steady-state analysis: (A) initial geometry of the sample and (B) hexahedral brick element meshing with finer mesh in the deformation zone (highlighted in *red*).

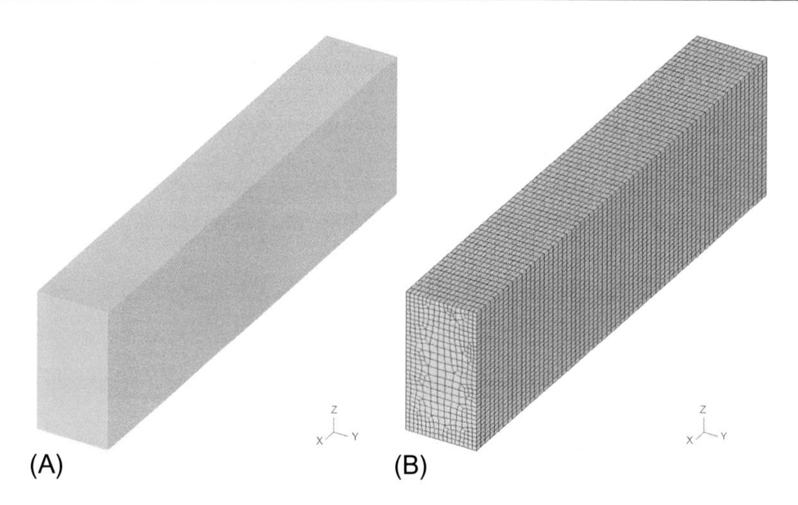

Fig. 4.5 Unsteady-state analysis: (A) initial geometry of the sample and (B) hexahedral brick element meshing.

complete picture of the process and also the material flow along the rolling and transverse directions, a full three-dimensional model has been considered. Also, to study the inhomogeneity in deformation one needs to consider the full geometry.

4.8.2 Mesh generation

The brick element and the tetrahedron element are the two types of elements normally used to discretize a three-dimensional object. It has been reported that brick element performs much better as compared with tetrahedron element. However, for objects with complex geometry, chances of geometry degradation are always there when discretized with brick elements. So, tetrahedron element is preferred for domains having complex geometry. Also the number of nodes involved in brick mesh is higher than the number of elements which increases the computational time. Furthermore, remeshing of the brick elements is difficult and hence may not be suitable for large deformation problems. But for problems without remeshing, brick elements are preferred (Li et al., 2001; Oh et al., 2001).

For the present study, as the workpiece is of simpler geometry and the amount of deformation involved is not too high, hexahedral brick elements have been used for the meshing of the workpiece. Simulations have been performed with different mesh sizes (i.e., different number of elements) to study the mesh sensitivity. The rolls are considered as rigid body and no meshing has been done.

4.8.2.1 Steady-state analysis

As the thermomechanical process takes place in the deformation zone only, mesh density with smaller element size in the deformation zone is used to get accurate results (Fig. 4.4B). But the ratio of the element size in the deformation zone and in other part of the sample should be optimum in order to achieve an efficient result. Higher mesh density in the deformation zone improves the simulation results in terms of accuracy and computational time. The nodes, in this approach (ALE), are fixed in the rolling direction and are updated in the transverse direction (direction perpendicular to rolling direction) (Li et al., 2009). So, the extreme boundary planes in the rolling direction, defined as entry surface and exit surface (Fig. 4.7C), do not move in the rolling direction, and hence do not capture the material spread along the lagging edge and leading edge, respectively. But the updated nodes along the transverse direction help in predicting the lateral spread of the material.

4.8.2.2 Unsteady-state analysis

Hexahedral brick element, with uniform element size along the length, has been used (Fig. 4.5B). However, a movable mesh window can be used in order to get smaller size elements in the deformation zone. The wireframe models of the workpiece and roll for the steady- and unsteady-state analyses are shown in Figs. 4.6A and B, respectively.

4.8.3 Mesh update

In ALE approach, initially the nodes, i.e., the computational reference points, are superimposed with the material points, like Lagrangian approach. With the increment in time, as the mesh deforms and changes its geometry, the new coordinates as well as the deformation state variables are obtained through computation. These values are then transferred to the material points to update the material reference system, but the computational reference system remains unchanged. Once the material points are updated, the computational nodes are updated to get a mesh with boundary coinciding with the boundary of the updated material points (Li et al., 2009). It is to be noted that with this update the position of the nodes along the rolling direction does not change. In Lagrangian approach, as the nodes are the integral part of the deforming material, there is no such complication in updating the mesh. Also as the nodes belong to the material points, no interpolation of the state variables is necessary.

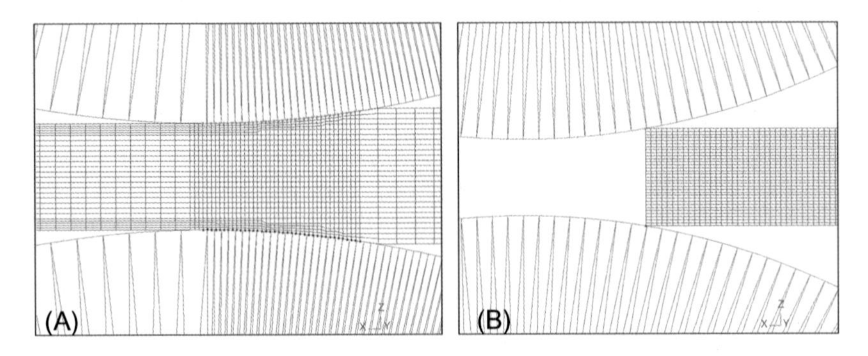

Fig. 4.6 Wireframe model of roll and workpiece for (A) steady-state analysis and (B) unsteady-state analysis.

4.8.4 Boundary conditions

4.8.4.1 Thermal boundary

Temperature is one of the most significant parameters during hot rolling process. Heat exchange between the workpiece and environment, including work-roll, is a complex process, which includes all the three modes of heat transfer. The thermal boundaries involved in the hot rolling process can be given as follows:

Boundary condition for cooling of workpiece in air

$$q_t = h_\infty (T_s - T_\infty) + \varepsilon\sigma\left(T_s^4 - T_\infty^4\right) \tag{4.4}$$

In the deformation zone,

$$q = h(T_s - T_\infty) \tag{4.5}$$

The value of IHTC, h, is critical for minimizing the error in the predicted FE solution as it directly controls the temperature of the workpiece, and hence the deformation stress. Reported values show a wide range, varying from 2 to 200 kW/m^2 K (Sellars, 1985). So, a few preliminary simulations have been carried out with different h values starting from 5 to 25 kW/m^2 K to match experimental rolling temperatures. Finally, a constant value of 15 kW/m^2 K has been used for all the passes. Convective heat transfer coefficient is taken as 0.02 kW/m^2 K, which is in the range of previously reported values (Phaniraj et al., 2005).

4.8.4.2 Mechanical boundary

The mechanical interaction of the workpiece with the work-roll is described by the shear type of interfacial friction.

$$\tau = mk_1 \tag{4.6}$$

where m is the friction factor and is taken as 0.7 (no lubricant is used) and k_1 is the shear yield strength of the workpiece material (Kobayashi et al., 1989).

The earlier mentioned boundary conditions for steady- and unsteady-state analyses are shown graphically in Figs. 4.7 and 4.8, respectively. For steady-state analysis, except for entry and exit surfaces, heat transfer from all other surfaces of the workpiece is considered, whereas for unsteady-state analysis heat transfer from all the surfaces is considered. Similarly, the initial mechanical boundary conditions have been defined at the roll-workpiece interface for steady-state analysis, and only at the top and bottom surfaces of the leading edge of the workpiece, which is the contact zone, for the unsteady-state analysis.

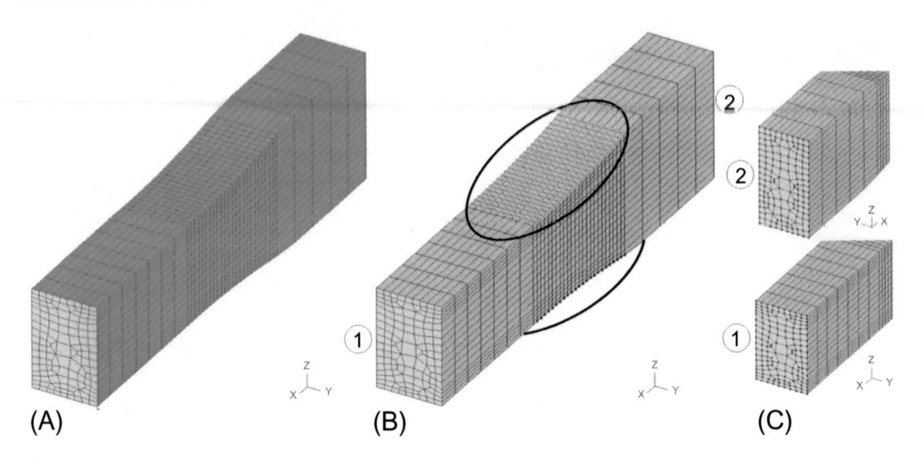

Fig. 4.7 Steady-state analysis: (A) thermal boundary conditions (highlighted in *green*), (B) initial contact boundary conditions, and (C) 1—exit surface and 2—entry surface.

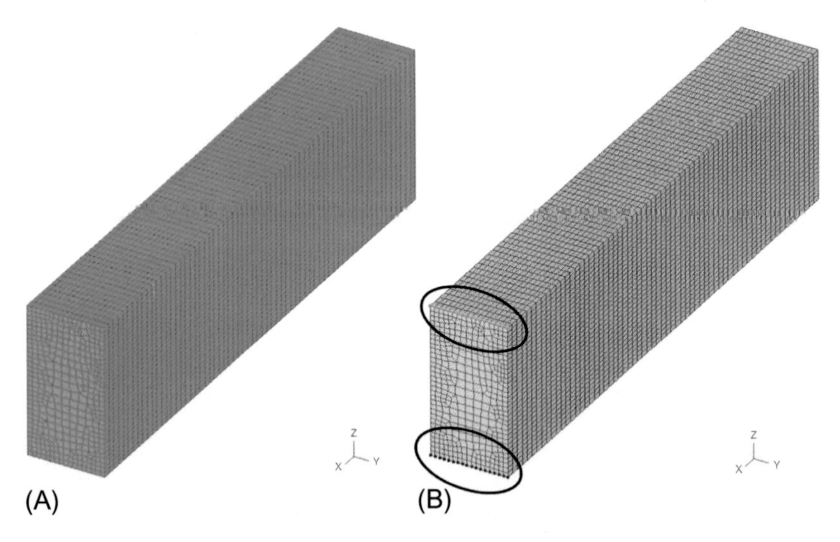

Fig. 4.8 Unsteady-state analysis: (A) thermal boundary conditions and (B) initial contact boundary conditions.

4.8.5 Governing equation

4.8.5.1 Equilibrium equations

The governing equation for equilibrium of the workpiece is given by

$$\frac{\partial \sigma_{ij}}{\partial x_j} = 0 \tag{4.7}$$

where σ_{ij} is the Cauchy stress tensor with $i,j = 1$, 2, and 3.

4.8.5.2 Constitutive equation

The analysis of metal forming processes is based on flow formulation considering infinitesimal deformation theory. The constitutive equation for the rigid-viscoplastic workpiece model is based on the Levy-Mises flow rule and can be written as (Kobayashi et al., 1989; Rout et al., 2016b):

$$\sigma'_{ij} = \frac{1}{\dot{\lambda}} \dot{\varepsilon}_{ij} \tag{4.8}$$

where $\sigma_{ij}{}'$ and $\dot{\varepsilon}_{ij}$ are the components of the deviatoric stress and strain rate tensor, respectively.

The coefficient $\dot{\lambda}$ is a function of state variables and can be expressed by (Kobayashi et al., 1989; Rout et al., 2016b):

$$\dot{\lambda} = \frac{3\dot{\overline{\varepsilon}}}{2\overline{\sigma}} \tag{4.9}$$

where $\dot{\overline{\varepsilon}}$ is the effective strain rate and $\overline{\sigma}$ is the effective stress.

$$\dot{\overline{\varepsilon}} = \sqrt{\frac{2}{3} \dot{\varepsilon}_{ij} \dot{\varepsilon}_{ij}} \tag{4.10}$$

The compatibility of the deforming work material is based on (Kobayashi et al., 1989; Rout et al., 2016b):

$$\dot{\varepsilon}_{ij} = \frac{1}{2} \left(\frac{\partial u_i}{\partial x_j} + \frac{\partial u_j}{\partial x_i} \right) \tag{4.11}$$

4.8.5.3 Yield function

The yielding of the material can be expressed by the von Mises' yield criteria and can be written as follows (Kobayashi et al., 1989; Rout et al., 2016b):

$$\overline{\sigma} = \sqrt{\frac{3}{2} \sigma'_{ij} \sigma'_{ij}} \tag{4.12}$$

4.8.6 Conservation laws

The behavior of the body undergoing deformation can be described by two sets of governing equations: the conservations laws and the constitutive equations.

The conservations laws are applicable to any material, whereas the constitutive equations differentiate the deformable bodies from material perspective.

4.8.6.1 Lagrangian formulation

The conservation laws in Lagrangian description can be expressed as (Priyadarshini et al., 2012):

Conservation of mass:

$$\dot{\rho} + \rho \operatorname{div} \vec{v} = 0 \tag{4.13}$$

Conservation of momentum:

$$\rho \dot{\vec{v}} = \vec{f} + \operatorname{div} \sigma \tag{4.14}$$

Conservation of energy:

$$\rho \dot{e} = \sigma : D - \operatorname{div} \vec{q} + q \tag{4.15}$$

where the superposed "•" indicates material derivatives in Lagrangian description, ρ is the mass density, $\vec{v}$ is the material velocity, $\vec{f}$ is the body forces, σ is the Cauchy stress tensor, e is the specific internal energy, D is the strain tensor, $\vec{q}$ is the heat flux vector, and q is the heat generation rate. The symbol ":" denotes the repetitive indices as $\sigma : D = \sigma_{ij} D_{ij}$.

4.8.6.2 ALE formulation

In ALE formulation, the earlier mentioned conservations laws (Eqs. 4.13–4.15) can be written as (Priyadarshini et al., 2012):

$$\tilde{\rho} + \vec{c}\, \nabla \rho + \rho \operatorname{div} \vec{v} = 0 \tag{4.16}$$

$$\rho \tilde{\vec{v}} + \rho\, \vec{c}\, \nabla \vec{v} = \vec{f} + \operatorname{div} \sigma \tag{4.17}$$

$$\rho \tilde{e} + \rho\, \vec{c}\, \nabla e = \sigma : D - \operatorname{div} \vec{q} + q \tag{4.18}$$

where the superposed "~" is defined in the ALE description, ∇ is the gradient operator, and $\vec{c}$ is the convective velocity defined as

$$\vec{c} = \vec{v} - \vec{\hat{v}} \tag{4.19}$$

where $\vec{v}$ is the material velocity and $\vec{\hat{v}}$ is the nodal velocity.

4.8.7 Constitutive equation for material model

Constitutive equation based on the Zener-Hollomon parameter (McQueen and Ryan, 2002; Mirzadeh et al., 2013) was used in the present study.

$$\dot{\bar{\varepsilon}} = A[\sinh(\alpha\bar{\sigma})]^n \exp\left(\frac{-Q}{RT}\right) \tag{4.20}$$

where A, α, and n are the material constants, Q is the activation energy, R is the gas constant, and T is the deformation temperature in Kelvin. The flow stress behavior of the material under consideration is dependent on strain, strain rate, and temperature, and hence the material is characterized as a rigid-viscoplastic material.

The material parameters were determined by performing a series of constant strain rate compression tests at 900°C, 1000°C, and 1100°C at strain rates of 0.01, 0.1, and 1 s^{-1}. The material constants for 304 austenitic stainless steel (the material under study) obtained from the tests listed in Table 4.4 were used.

4.8.8 Finite element formulation

4.8.8.1 Deformation analysis

The formulation of finite element equations for an element can be derived by three different approaches: (i) direct approach, (ii) variational approach, and (iii) weighted residual approach. The direct approach, as its name indicates, can directly relate to the direct stiffness method of structural analysis. This method needs a knowledge of matrix algebra and is normally applicable to simple problems involving simple element shapes, whereas the variational approach is based on the calculus of variations and includes extremizing a functional. The functional may be the complementary energy, the potential energy, or any alternative to these energies. The weighted residual approach is more versatile and does not rely on variations. This method is widely used in fluid mechanics and heat transfer problems (Huebner et al., 2004). The metal forming processes usually involve nonlinear equations due to which the calculation of these processes becomes complicated. Based on this, the variational approach can be the best approach in solving the metal forming processes through FEM. Because the matrices formed by the variational approach are always symmetric it makes the computation easier (Rahman and Agrawal, 2013).

Table 4.4 Values for material constants for 304 austenitic stainless steel

Parameter	$A\ (s^{-1})$	$\alpha\ (MPa^{-1})$	n	Q (kJ/mol)
Value	1×10^{18}	6.4×10^{-3}	5.3	473

The FE formulation for the rigid-viscoplastic material, based on the variational approach, depicts among all the admissible velocities u_i the actual velocities which satisfy the compatibility conditions, the incompressibility conditions, and the velocity boundary conditions and gives the following functional a stationary value (Kobayashi et al., 1989):

$$\pi = \int_V E(\dot{\varepsilon}_{ij})dV - \int_{S_v} F_i u_i dS \tag{4.21}$$

where $E(\dot{\varepsilon}_{ij})$ denotes the work function and F_i is the traction force. A penalized form of the incompressibility is added to remove the incompressibility constraint on admissible velocity fields. The actual velocity field can now be determined from the stationary value of the variation as follows,

$$\delta\pi = \int_V \overline{\sigma}\delta\dot{\overline{\varepsilon}}dV + K_1 \int_V \dot{\varepsilon}_v \delta\dot{\varepsilon}_v dV - \int_{S_F} F_i \delta u_i dS = 0 \tag{4.22}$$

where $\overline{\sigma} = \overline{\sigma}(\overline{\varepsilon}, \dot{\overline{\varepsilon}})$ represents the rigid-viscoplastic materials and $\dot{\varepsilon}_v = \dot{\varepsilon}_{ii}$ is the volumetric strain rate. However, when $\overline{\sigma} = \overline{\sigma}(\overline{\varepsilon})$, the previous equation can be applicable for rigid-plastic materials. The penalty constant, K_1, should be a very large positive constant. The solution to Eq. (4.22) can be obtained from the admissible velocity fields which are defined with the help of velocity field of each element. In order to maintain the continuity in the velocity field, shape functions were defined. The defined shape functions maintain the compatibility across the elements.

Eq. (4.22) can be expressed in terms of nodal point velocities, v and their variations δv. The set of algebraic equations, i.e., stiffness equations are obtained from arbitrariness of δv_I as (Kobayashi et al., 1989):

$$\frac{\partial \pi}{\partial v_I} = \sum_j \left(\frac{\partial \pi}{\partial v_I}\right)_j = 0 \tag{4.23}$$

where I represents the nodal point number and j indicates the number of element. This nonlinear stiffness equation is linearized by Taylor expansion for an initial guess $v = v_o$ such that (Kobayashi et al., 1989)

$$\left[\frac{\partial \pi}{\partial v_I}\right]_{v=v_0} + \left[\frac{\partial^2 \pi}{\partial v_I \partial v_J}\right]_{v=v_0} \Delta v_J = 0 \tag{4.24}$$

where Δv_J is the first-order correction for v_0. The previous equation can also be written as (Kobayashi et al., 1989)

$$K\Delta v = f \tag{4.25}$$

where K is the stiffness matrix and f is the residual nodal point force vector. The equation is solved iteratively until the velocity correction term, Δv and the residual of force vector converge. Once the solution converges, workpiece geometry is updated and the computation is repeated until the desired deformation is achieved.

4.8.8.2 Thermal analysis

The temperature distribution in the workpiece can be calculated by using the governing partial differential equation of heat transfer:

$$k\nabla^2 T + \dot{q} = \rho C \dot{T} \tag{4.26}$$

Here, $\dot{q}$, the rate of heat generated during the rolling process, consists of both heat generated due to friction ($\dot{q}_f$) and due to plastic deformation ($\dot{q}_m$). The mechanical energy per unit volume spent in deformation is equal to the area under the stress-strain curve, and the heat generated due to the same can be calculated by using equation (Bagheripoor and Bisadi, 2011):

$$\dot{q}_m = \hat{n}\bar{\sigma}\dot{\bar{\varepsilon}} \tag{4.27}$$

The model also considers the heat generated due to friction which can be defined as,

$$\dot{q}_f = \tau |v| \tag{4.28}$$

For arbitrary variation in temperature δT, Eq. (4.26) can be expressed in the form (Kobayashi et al., 1989):

$$\int_V k\nabla^2 T\, \delta T\, dV + \int_V \hat{n}\bar{\sigma}\dot{\bar{\varepsilon}}\delta T\, dV - \int_V \rho C\dot{T}\, \delta T\, dV = 0 \tag{4.29}$$

In FE formulation, the temperature field in the previous equation can be approximated by

$$T = \sum_{\alpha} q_\alpha T_\alpha = N^T T \tag{4.30}$$

where q_α is the shape function and T_α is the temperature at αth node.

The previous equation can be expressed in the following form (details can be found in Kobayashi et al., 1989):

$$C\dot{T} + K_C T = Q \tag{4.31}$$

where C is the heat capacity matrix, $\dot{T}$ is the nodal point temperature rates vector, K_C is the heat conduction matrix, T is the nodal point temperature vector, and Q is the heat flux vector. By implementing the boundary conditions, Eq. (4.31) can be expressed, with an interpolation function N, as (Kobayashi et al., 1989):

$$Q = \int_V \hat{n}\bar{\sigma}\dot{\bar{\varepsilon}} N\,dV + \int_S \varepsilon\sigma\left(T_s^4 - T_\infty^4\right) N\,dS + \int_S h_\infty (T_s - T_\infty) N\,dS + \int_S \dot{q}_f N\,dS \qquad (4.32)$$

4.8.9 Type of solver

Numerical simulations of the forming process are performed incrementally and in each time increment, a set of nonlinear equations, i.e., stiffness equations is solved. The nonlinear equations are first converted to linear equations. Linear equations are obtained by Taylor expansion with an initial guess to the solution. These linear equations are then solved either by direct method or by iterative method. Conjugate-gradient (CG) method is the most popular iterative solver, whereas in direct method sparse solver is usually used. The direct solver usually takes more time to solve the equations and requires more memory space as compared to iterative methods. Also, the iterative methods are easier to implement for parallel computation. However, as compared to iterative methods, the convergence rate is faster in direct methods. Details about the earlier mentioned solvers can be found elsewhere (Saad, 2003). At each time step of computation, better solution to the linear equations is achieved iteratively either by Newton-Raphson or by direct iteration method.

For the present study, the problem has been considered as a coupled thermomechanical problem. So, simulations are carried out to solve both the deformation and temperature equations, simultaneously. The equations for the deformation analysis converge when the velocity and force error limits are satisfied (Kobayashi et al., 1989); and for this study these values are 0.005 and 0.05, respectively. It is to be noted that the computation time can be reduced by increasing these values. However, at the same time this may increase the error between simulation and the experimental results. Maximum number of iterations for each time step has been set as 200. For steady-state analysis the simulations have been performed until the steady-state condition has been reached, whereas for the unsteady-state analysis, simulations were run till the workpiece comes out of the roll gap. The process has reached the steady-state condition in ~0.422 s. However, the exact time to reach the steady-state condition varies with number of total elements.

The equations for deformation and thermal analyses are coupled and solved simultaneously. Solving the earlier mentioned thermo-viscoplastic problem through finite element procedure is summarized in Fig. 4.9 (Kobayashi et al., 1989).

4.8.10 Computing environment

In FEM, the time required to perform the simulation not only depends on the number of elements and complexity of the problem, but also depends on the device on which the computations are being performed. However, the availability of the high

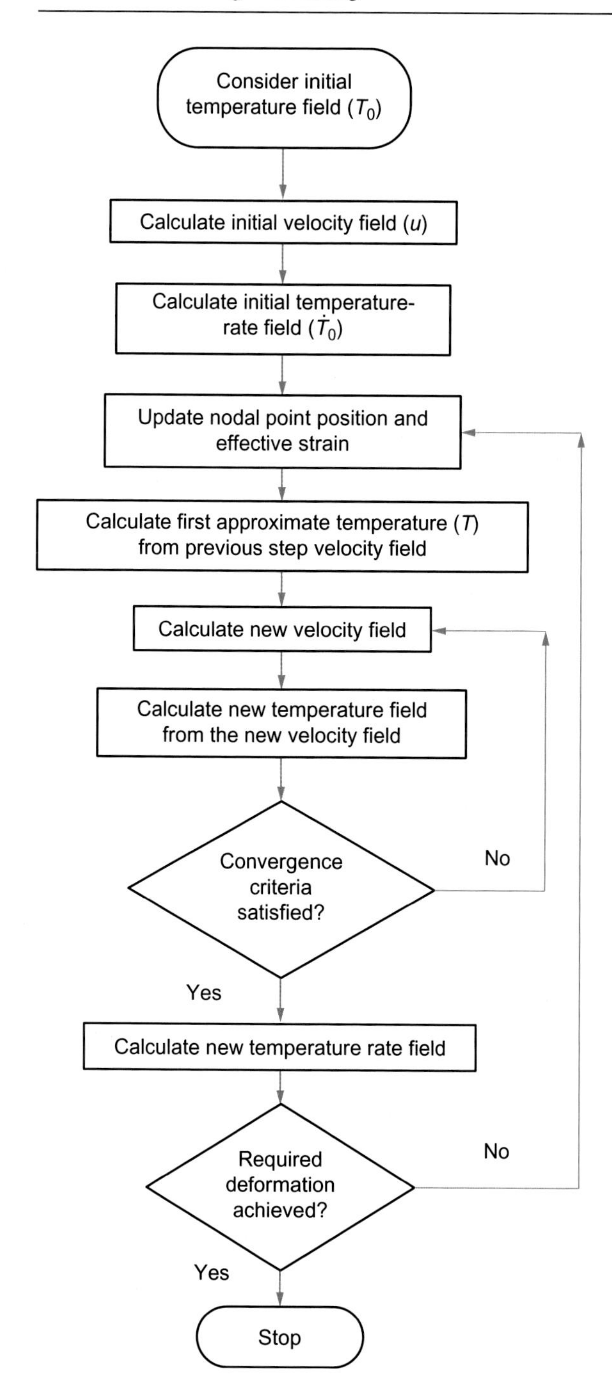

Fig. 4.9 Flowchart of the procedure to solve thermo-viscoplastic problem through FE.

configured computers has reduced the computational time significantly. This is also one of the reasons for the popularity of FEM in the field of metal forming processes. The present simulations were performed on a 64-bit computer configured with Intel core i5, 3.2 GHz processor with 4 GB RAM. As expected, a longer computational time is taken for the mesh with smaller element size.

4.8.11 Mesh sensitivity study

Simulations have been performed to study the mesh sensitivity for a single pass hot rolling. Details of the process parameters and the FE mesh on which the simulations were performed are presented in Table 4.5. For steady-state analysis, the element size within the deformation zone has been mentioned. Simulations were run on both the CG and sparse solvers. To compare the effect of mesh sensitivity on the simulation results, rolling load and workpiece temperature have been considered. It has been found that, for the present case, rolling load and surface temperature are not much sensitive to the number of elements. Surface temperature of the workpiece in the deformation zone depends directly on interfacial heat transfer coefficient, h, which is same for all the simulation throughout the mesh sensitivity analysis. Thus there is not much difference in surface temperature with increase in the number of elements. Similarly, for rolling force also, no significant deviation has been observed.

Fig. 4.10A shows the simulation time taken by the sparse and the CG solvers for steady- and unsteady-state analyses. In all the cases, as expected, the computational time increases with the increase in number of elements. However, the time taken by sparse solver is higher than that taken by the CG solver. It can be noted that the simulation for the unsteady-state analysis has been run for a longer processing time (to complete rolling process) which adds to the higher computational time. The same reason can also be applicable for the memory consumption as it can be clearly observed from Fig. 4.10B that the memory consumed by the unsteady-state analysis is higher than that for the steady-state analysis. However, the memories occupied by the two different solvers are almost the same for steady- and unsteady-state analyses.

In the following sections, comparison of results for both the analyses has been made. For this, the results of the simulation carried out for a total element of about 25,000 (SS4 and US4) and with sparse solver have been considered. In steady-state analysis, the steady state has been achieved at the processing time of 0.43 s. To make a comparison of the results from the two different analyses, same processing time has been considered, i.e., at time $t = 0.43$ s.

4.8.12 Temperature distribution

The temperature distributions in the workpiece for the steady- and unsteady-state analyses are shown in Figs. 4.11A and B, respectively. Both the analyses predicted the same trend in temperature distribution. In both the cases, a much lower surface temperature has been observed in the contact zone, whereas temperature higher than initial sample temperature of the workpiece has been observed at the center of the

Table 4.5 **Mesh details for mesh sensitivity study**

Name	Roll dimension: 160 mm radius, 300 mm width Roll speed: 10 rpm				Initial temperature: 1000 °C	Sample dimension: 11 × 20.4 × 78 mm	
	Reduction: 4 mm (single pass)				IHTC: 15 kW/m^2 °C	Material: 304 SS	
	Number of elements (hexahedral brick)						
Name	**Cross section**	**Along length**	**Total**	**No. of nodes**	**Maximum size of element (mm)**	**Minimum size of element (mm)**	**Ratio of element size**
	Steady-state analysis (element size in the deformation zone)						
SS1	130	36	4680	5809	2.44281	0.430027	5.681
SS2	184	48	8832	10,633	2.10533	0.353025	5.964
SS3	271	60	16,260	19,032	1.4944	0.292595	5.107
SS4	352	72	25,344	29,200	1.39603	0.218175	6.398
	Unsteady-state analysis						
US1	130	42	5460	6751	2.44283	0.478268	5.107
US2	228	50	11,400	13,566	1.5451	0.542855	2.846
US3	309	58	17,922	21,358	1.484	0.414756	3.578
US4	389	65	25,285	28,974	1.32316	0.361955	3.655

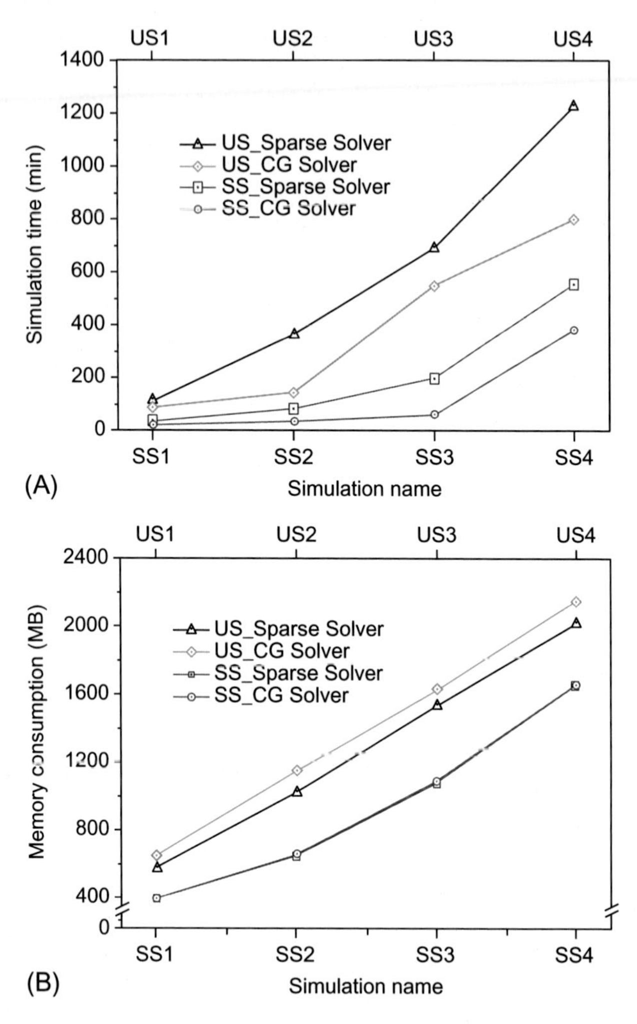

Fig. 4.10 (A) Simulation time taken and (B) memory consumed by sparse and CG solvers for steady- and unsteady-state analyses.

workpiece. No significant difference in temperature values has been observed for the two different analyses.

As in unsteady-state analysis, the process is simulated from the beginning to the end and the formulation being Lagrangian, it is possible to get the temperature variation at different material points throughout the process. Different points near to the workpiece surface have been defined to track the temperature variation. The defined points are shown in Fig. 4.12A and the temperature variation is shown in Fig. 4.12B.

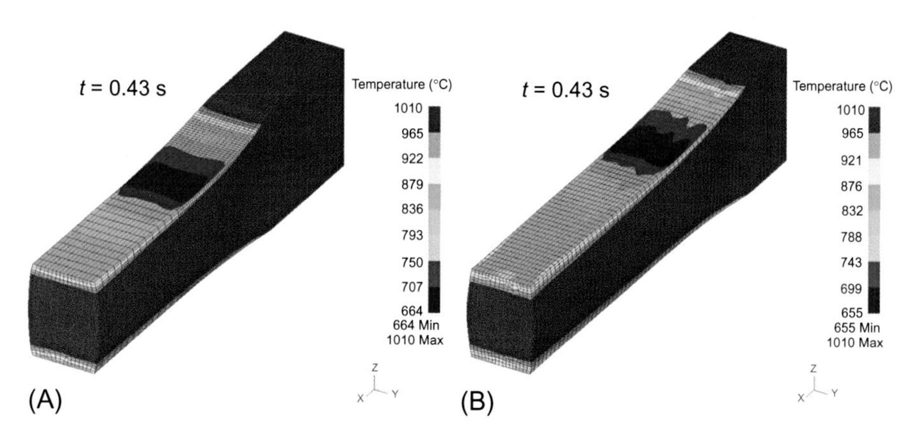

Fig. 4.11 Temperature distributions on the workpiece at time $t = 0.43$s for (A) steady-state analysis and (B) unsteady-state analysis.

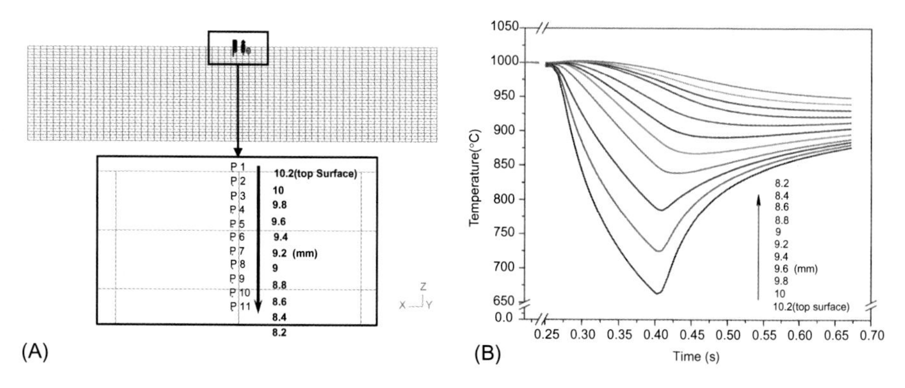

Fig. 4.12 (A) Points on the workpiece model to track temperature variation during the unsteady-state analysis and (B) temperature variation on the defined points.

The surface temperature of the workpiece decreases significantly, as it comes in contact with the work-roll, due to high IHTC. This phenomenon is termed as *roll chilling effect* (Chen et al., 1993; Krzyzanowski et al., 2010). However, the temperature value increases after rolling due to the conduction of heat from the center of the plate. The significant drop in temperature is up to 1 mm (1/20th of thickness) from the surface. The same trend is followed from surface toward the center of the workpiece. However near the center, the workpiece temperature is more than the initial temperature. This is due to the heat accumulation from plastic deformation. In other way, it can be said that the heat capacity of the plate material plays a significant role for the center or internal temperature variation, while IHTC plays an important role for surface temperature.

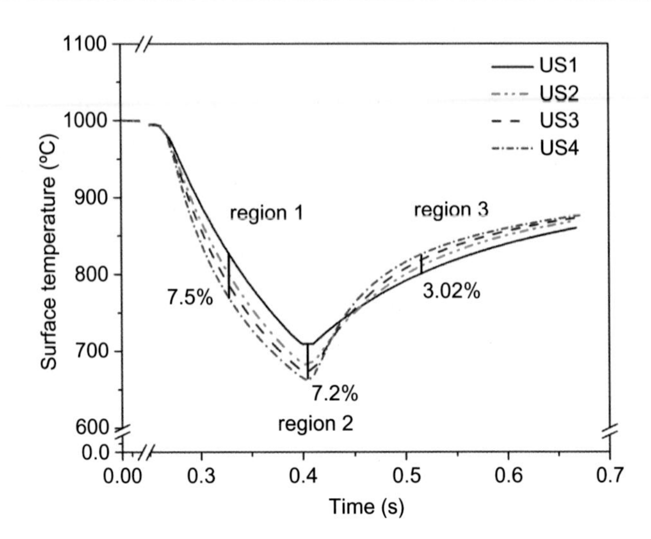

Fig. 4.13 Surface temperature variation of workpiece with different numbers of elements for unsteady-state analysis with sparse solver.

Fig. 4.13 shows the surface temperature variation (at point P1 of Fig. 4.12A) of the workpiece during the unsteady-state analysis. The variation is plotted for different simulation runs with different number of elements. For better understanding, the temperature variation plot can be compared among three different regions—region 1: decrease in surface temperature, region 2: minimum surface temperature, and region 3: rise in surface temperature. Decrease in surface temperature occurs when the selected point P1 comes in contact with the roll, i.e., during the rolling process, whereas rise in surface temperature occurs when P1 comes out of the deformation zone, i.e., after the rolling process. From Fig. 4.13 it can be observed that with increase in the number of element, both the surface temperature drop (region 1) and surface temperature rise (region 3) increase. The minimum surface temperature (region 2) predicted by the simulation, US1, is around 709°C, whereas it is 660°C for the simulation US4. So, a deviation of ~7.2% has been observed between the considered minimum and maximum numbers of mesh elements, i.e., US1 and US4, respectively. Similarly, the maximum variations in drop of surface temperature (region 1) and rise of surface temperature (region 3) were found to be ~7.5% and ~3.02%, respectively.

4.8.13 Stress and strain distributions

Figs. 4.14A and B show the effective stress distribution in the workpiece for the steady- and unsteady-state analyses, respectively. The predicted stress distribution, by both the analyses, is identical. A higher stress value is observed at the workpiece surface near the exit zone. The 304 austenitic SS material is sensitive to temperature and strain rate, as seen from the constitutive equation (Eq. 4.20). The flow stress value increases with the increase in strain rate, whereas it decreases with the increase in deformation temperature. So, the stress distribution in the workpiece is dependent

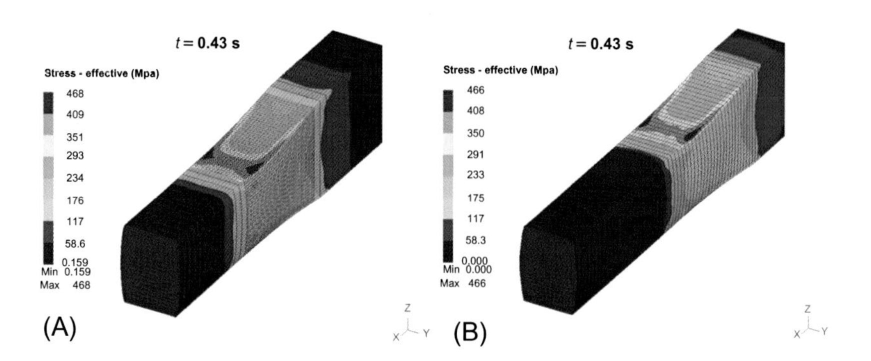

Fig. 4.14 Effective stress distribution on the workpiece at time $t = 0.43$ s for (A) steady-state analysis and (B) unsteady-state analysis.

on the temperature and strain rate distributions. Due to the longer contact time between the workpiece and the roll, the surface temperature value is low near the exit zone as compared with the entry zone (Figs. 4.11A and B). However, the strain rate value will be higher in the entry zone as compared with the exit zone. This is because of the immediate interaction of the workpiece with the roll causing a high local deformation near the entry zone. The stress distribution in the workpiece, showing a higher value at the surface near the exit zone, clearly indicates the predicted stress is very much sensitive to deformation temperature than the strain rate. Below the surface, as the temperature of the workpiece is higher, a moderate stress value has been observed. The considered material model is a rigid-viscoplastic, which does not consider the elastic strain. So, in both the analyses stress in the deformation zone is calculated.

Figs. 4.15A and B show the effective strain distributions on the workpiece for steady- and unsteady-state analyses, respectively. Inhomogeneity in strain along

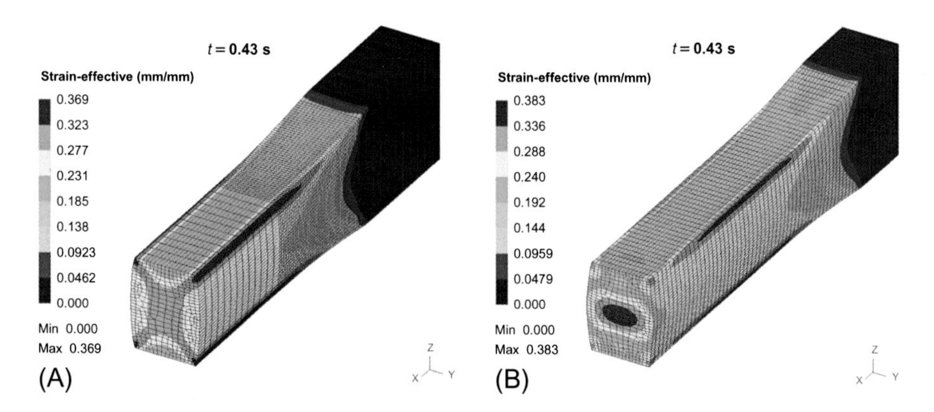

Fig. 4.15 Effective strain distribution on the workpiece at time $t = 0.43$ s for (A) steady-state analysis and (B) unsteady-state analysis.

the thickness direction can be observed in both the cases. A higher amount of strain has been developed at the center as compared with the surface of the workpiece, which is not normally experienced in case of industrial rolling. This could be due to the lower width of the workpiece so that the lateral spread at the center is causing a higher effective strain. Also, presence of friction at the surface restricts the lateral spread at the surface. Though the strain distribution in the deformation zone is same in both the analyses, a difference in the distribution has been observed on the rolled portion of the workpiece. At the surface, the steady-state analysis predicts a higher strain. This may be due to the larger element size as compared to that at the deformation zone. Also at the center of the leading edge, unsteady-state analysis predicts a higher strain value. This is possibly because the material at the leading edge is free to move along both the lateral direction and rolling direction, which might have increased the effective strain value. However, in case of steady-state analysis the movement of material, near the leading edge, along the rolling direction has not been considered. So, to study the strain distribution or material flow or edge profile in hot rolling, unsteady-state analysis is preferred. Also for analysis of defects like wavy edge, the total rolling process needs to be simulated for which unsteady-state approach is needed.

4.8.14 Rolling load and rolling torque

Accurate predictions of rolling load and rolling torque are very much essential in roll pass design. It is also important to know these parameters in order to prevent breakdown of rolling mill and to ensure maximum productivity without compromising the geometry and the desired properties of the workpiece. Figs. 4.16A and B show the predicted rolling load and rolling torque for the simulations with ~25,000 (SS4 and US4) elements, respectively, and computed with sparse solver. Rolling experiment on the samples of same dimensions with same process parameters has been performed to compare the predicted rolling load value with the experimentally obtained load value. For unsteady-state analysis, the rolling load value gradually increases as

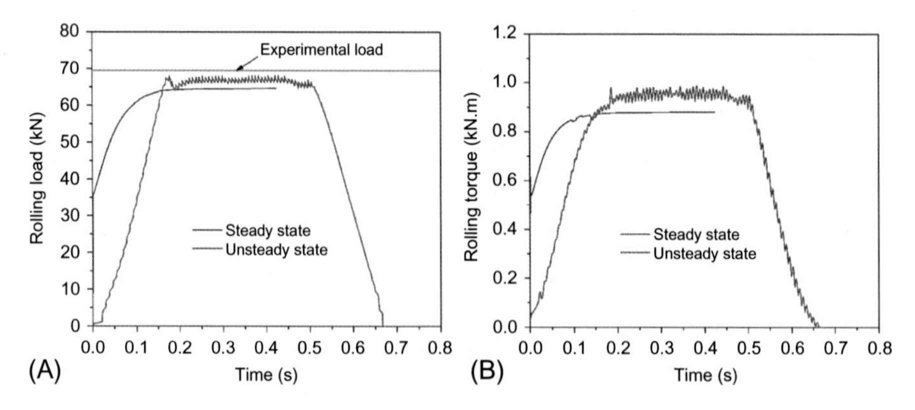

Fig. 4.16 Comparison of predicted rolling load and torque using steady- and unsteady-state analyses with experimentally measured values: (A) rolling load and (B) rolling torque.

the workpiece enters into the roll gap, attains a relatively steady value, and again starts to decrease when the workpiece comes out of the roll gap. The same trend is also followed for the rolling torque. However, for the steady-state analysis, as the whole process is not analyzed, the rolling load is only calculated till the arrival of the steady state. In both the analyses the average rolling load during the steady zone is considered as the rolling load. A good agreement between the experimental and predicted load and torque values can be observed.

4.9 Concluding remarks

The chapter gives an overview of the work carried out on FE modeling of the thermomechanical analysis of hot rolling process. It discussed the research works done since the implementation of FEM in metal forming process and mostly focuses on the thermomechanical analysis of the deformable workpiece. It also describes about the workpiece and roll interface behavior, i.e., the friction coefficient and the interfacial heat transfer coefficient. A brief description to the basics of FEM is also given. In the end, details on both kind of analyses of hot rolling, i.e., steady- and unsteady-state thermomechanical analyses of the three-dimensional workpiece have been presented.

The application of FE modeling to various engineering problems is increasing; however, the implementation of the same to hot rolling process is started in 1970s. Furthermore, improvisation to the FE modeling to increase the computational efficiency and also to tackle the various critical problems of the hot rolling process is still going on. The choice of analysis, i.e., steady state or unsteady state, purely depends on the output requirement of the analyzer. If the requirement is restricted to the energy consumption by predicting the rolling load and torque then, perhaps, steady-state analysis is the most preferred method. However, the unsteady-state analysis can be extended to many aspects, such as analysis of scale formation on workpiece surface, defects analysis, edge profile study, through-thickness and through-process temperature analyses, and hence microstructure evolution, etc. However, higher computational time is a matter of concern.

References

Aiyedun, P.O., Sparling, L.G.M., Sellars, C.M., 1997. Temperature changes in hot flat rolling of steels at low strain rates and low reduction. Proc. Inst. Mech. Eng. B J. Eng. Manuf. 211, 261–284. http://dx.doi.org/10.1243/0954405971516257.

Bagheripoor, M., Bisadi, H., 2011. Effects of rolling parameters on temperature distribution in the hot rolling of aluminum strips. Appl. Therm. Eng. 31, 1556–1565. http://dx.doi.org/10.1016/j.applthermaleng.2011.01.005.

Ben, Z., Jing, W., Heng-Hua, Z., 2011. Numerical simulation of multi-pass rolling force and temperature field of plate steel during hot rolling. J. Shanghai Jiaotong Univ. 16, 141–144. http://dx.doi.org/10.1007/s12204-011-1109-4.

Brand, A.J., Kalz, S., Kopp, R., 1996. Microstructural simulation in hot rolling of aluminium alloys. Comput. Mater. Sci. 7, 242–246. http://dx.doi.org/10.1016/S0927-0256(96)00087-0.

Byon, S.M., Kim, S.I., Lee, Y., 2004. Predictions of roll force under heavy-reduction hot rolling using a large-deformation constitutive model. Proc. Inst. Mech. Eng. B J. Eng. Manuf. 218, 483–494. http://dx.doi.org/10.1177/095440540421800502.

Byon, S.M., Na, D.H., Lee, Y.S., 2013. Flow stress equation in range of intermediate strain rates and high temperatures to predict roll force in four-pass continuous rod rolling. Trans. Nonferrous Met. Soc. China 23, 742–748. http://dx.doi.org/10.1016/S1003-6326(13)62524-8.

Cavaliere, M.A., Goldschmit, M.B., Dvorkin, E.N., 2001. Finite element simulation of the steel plates hot rolling process. Int. J. Numer. Methods Eng. 52, 1411–1430. http://dx.doi.org/10.1002/nme.262.

Chen, B.K., Thomson, P.F., Choi, S.K., 1992. Temperature distribution in the roll-gap during hot flat rolling. J. Mater. Process. Technol. 30, 115–130. http://dx.doi.org/10.1016/0924-0136(92)90042-Q.

Chen, W.C., Samarasekera, I.V., Hawbolt, E.B., 1993. Fundamental phenomena governing heat transfer during rolling. Metall. Trans. A. 24A, 1307–1320. http://dx.doi.org/10.1007/BF02668199.

Chenot, J.L., Massoni, E., 2006. Finite element modelling and control of new metal forming processes. Int. J. Mach. Tools Manuf. 46, 1194–1200. http://dx.doi.org/10.1016/j.ijmachtools.2006.01.031.

Dawson, P.R., 1978. Viscoplastic finite element analysis of steady-state forming processes including strain history and stress flux dependence. Appl. Numer. Meth. Form. Process. 28, 55–66. ASME, AMD.

Devadas, C., Samarasekera, I.V., Hawbolt, E.B., 1991. The thermal and metallurgical state of steel strip during hot rolling: Part III. Microstructural evolution. Metall. Trans. A. 22A, 335–349. http://dx.doi.org/10.1007/BF02656800.

Ding, H., Hirai, K., Homma, T., Kamado, S., 2010. Numerical simulation for microstructure evolution in AM50 Mg alloy during hot rolling. Comput. Mater. Sci. 47, 919–925. http://dx.doi.org/10.1016/j.commatsci.2009.11.024.

Ding, H.L., Wang, T.Y., Yang, L., Kamado, S., 2013. FEM modeling of dynamical recrystallization during multi-pass hot rolling of AM50 alloy and experimental verification. Trans. Nonferrous Met. Soc. China 23, 2678–2685. http://dx.doi.org/10.1016/S1003-6326(13)62784-3.

Ding, Y., Zhu, Q., Le, Q., Zhang, Z., Bao, L., Cui, J., 2015. Analysis of temperature distribution in the hot plate rolling of Mg alloy by experiment and finite element method. J. Mater. Process. Technol. 225, 286–294. http://dx.doi.org/10.1016/j.jmatprotec.2015.06.011.

Duan, X., Sheppard, T., 2002. Three dimensional thermal mechanical coupled simulation during hot rolling of aluminium alloy 3003. Int. J. Mech. Sci. 44, 2155–2172. http://dx.doi.org/10.1016/S0020-7403(02)00164-9.

Duan, X., Sheppard, T., 2004. The influence of the constitutive equation on the simulation of a hot rolling process. J. Mater. Process. Technol. 150, 100–106. http://dx.doi.org/10.1016/j.jmatprotec.2004.01.026.

Dvorkin, E.N., Goldschmit, M.B., Cavaliere, M.A., Amenta, P.M., Marini, O., Stroppiana, W., 1997. 2D finite element parametric studies of the flat-rolling process. J. Mater. Process. Technol. 68, 99–107. http://dx.doi.org/10.1016/S0924-0136(96)00027-1.

Esteban, L., Elizalde, M.R., Ocaña, I., 2007. Mechanical characterization and finite element modelling of lateral spread in rolling of low carbon steels. J. Mater. Process. Technol. 183, 390–398. http://dx.doi.org/10.1016/j.jmatprotec.2006.10.047.

Galantucci, L.M., Tricarico, L., 1999. Thermo-mechanical simulation of a rolling process with an FEM approach. J. Mater. Process. Technol. 92–93, 494–501. http://dx.doi.org/10.1016/S0924-0136(99)00242-3.

Hu, P., Ying, L., Li, Y., Liao, Z., 2013. Effect of oxide scale on temperature-dependent interfacial heat transfer in hot stamping process. J. Mater. Process. Technol. 213, 1475–1483. http://dx.doi.org/10.1016/j.jmatprotec.2013.03.010.

Huang, C.Q., Deng, H., Diao, J.P., Hu, X.H., 2011. Numerical simulation of aluminum alloy hot rolling using DEFORM-3D. In: Proc. - 2011 IEEE Int. Conf. Comput. Sci. Autom. Eng. CSAE 2011, vol. 3, pp. 378–381. http://dx.doi.org/10.1109/CSAE.2011.5952701.

Huebner, K.H., Dewhirst, D.L., Smith, D.E., Byrom, T.G., 2004. The Finite Element Method for Engineers, fourth ed. John Wiley & Sons, Inc, Singapore.

Hutton, D.V., 2004. Fundamentals of Finite Element Analysis, first ed. The McGraw Hill Companies, New York. 10.1017/CBO9781107415324.004.

Jain, R., Kumari, K., Kesharwani, R.K., Kumar, S., Pal, S.K., Singh, S.B., Panda, S.K., Samantaray, A.K., 2015. Friction stir welding: scope and recent development. In: Davim, J.P. (Ed.), Mod. Manuf. Eng. Mater. Forming, Mach. Tribol. Springer International Publishing, Switzerland, pp. 179–229. http://dx.doi.org/10.1007/978-3-319-20152-8.

Kanazawa, K., Marcal, P.V., 1978. Finite element analysis on the steel rolling process. Appl. Numer. Meth. Form. Process. ASME, AMD 28, 81.

Kim, N., Kobayashi, S., 1990. Three-dimensional simulation of gap-controlled plate rolling by the finite element method. Int. J. Mach. Tools Manuf. 30, 269–281.

Kiuchi, M., Yanagimoto, J., Wakamatsu, E., 2000. Overall thermal analysis of hot plate/sheet rolling. CIRP Ann. Manuf. Technol. 49, 209–212. http://dx.doi.org/10.1016/S0007-8506(07)62930-8.

Kobayashi, S., 1982. A review on the finite-element method and metal forming process modeling. J. Appl. Metalwork. 2, 163–169. http://dx.doi.org/10.1007/BF02834034.

Kobayashi, S., Oh, S.-I., Altan, T., 1989. Metal Forming and the Finite Element Method. Oxford University Press, New York.

Krzyzanowski, M., Beynon, J.H., Sellars, C.M., 2000. Analysis of secondary oxide-scale failure at entry into the roll gap. Metall. Mater. Trans. B 31B, 1483–1490. http://dx.doi.org/10.1007/s11663-000-0033-z.

Krzyzanowski, M., Beynon, J.H., Farrugia, D.C.J., 2010. Oxide Scale Behaviour in High Temperature Metal Processing. Wiley-VCH Verlag GmbH & Co. KGaA, Weinheim. http://www.worldcat.org/oclc/477295377.

Lau, A.C.W., Shivpuri, R., Chou, P.C., 1989. An explicit time integration elastic-plastic finite element algorithm for analysis of high speed rolling. Int. J. Mech. Sci. 31, 483–497. http://dx.doi.org/10.1016/0020-7403(89)90098-2.

Lee, Y., 2002. New approach for prediction of roll force in rod rolling. Ironmak. Steelmak. 29, 459–468. http://dx.doi.org/10.1179/030192302225004647.

Li, G.-J., Kobayashi, S., 1982. Rigid-plastic finite-element analysis of plane strain rolling. J. Eng. Ind. 104, 55–63. http://dx.doi.org/10.1115/1.3185797.

Li, Y.H., Sellars, C.M., 1998. Comparative investigations of interfacial heat transfer behaviour during hot forging and rolling of steel with oxide scale formation. J. Mater. Process. Technol. 80–81, 282–286. http://dx.doi.org/10.1016/S0924-0136(98)00112-5.

Li, G., Jinn, J.T., Wu, W.T., Oh, S.I., 2001. Recent development and applications of three-dimensional finite element modeling in bulk forming processes. J. Mater. Process. Technol. 113, 40–45. http://dx.doi.org/10.1016/S0924-0136(01)00590-8.

Li, G., Yang, J., Oh, J.Y., Foster, M., 2009. Advancements of extrusion simulation in DEFORM-3D. Light Met. Age, 1–5.

Lin, Z., Shen, C., 1997. A rolling process two-dimensional finite element model analysis. Finite Elem. Anal. Des. 26, 143–160.

Lindgren, L.-E., Edberg, J., 1990. Explicit versus implicit finite element formulation in simulation of rolling. J. Mater. Process. Technol. 24, 85–94.

Liu, C., Hartley, P., Sturgess, C.E.N., Rowe, G.W., 1987. Finite-element modelling of deformation and spread in slab rolling. Int. J. Mech. Sci. 29, 271–283. http://dx.doi.org/10.1016/0020-7403(87)90040-3.

McQueen, H.J., Ryan, N.D., 2002. Constitutive analysis in hot working. Mater. Sci. Eng. A 322, 43–63. http://dx.doi.org/10.1016/S0921-5093(01)01117-0.

Min ting, W., Xin liang, Z., Xue tong, L., Feng shan, D., 2007. Finite element simulation of hot strip continuous rolling process coupling microstructural evolution. J. Iron Steel Res. Int. 14, 30–36. http://dx.doi.org/10.1016/S1006-706X(07)60039-9.

Mirzadeh, H., Parsa, M.H., Ohadi, D., 2013. Hot deformation behavior of austenitic stainless steel for a wide range of initial grain size. Mater. Sci. Eng. A 569, 54–60. http://dx.doi.org/10.1016/j.msea.2013.01.050.

Montmitonnet, P., 2006. Hot and cold strip rolling processes. Comput. Methods Appl. Mech. Eng. 195, 6604–6625. http://dx.doi.org/10.1016/j.cma.2005.10.014.

Montmitonnet, P., Chenot, J.L., Bertrand-Corsini, C., David, C., Iung, T., Buessler, P., 1992. A coupled thermomechanical approach for hot rolling by a 3D finite element method. J. Eng. Ind. Trans. ASME 114, 336–345.

Montmitonnet, P., Gratacos, P., Ducloux, R., 1996. Application of anisotropic viscoplastic behaviour in 3D finite-element simulations of hot rolling. J. Mater. Process. Technol. 58, 201–211. http://dx.doi.org/10.1016/0924-0136(95)02095-0.

Mori, K., Osakada, K., 1984. Simulation of three-dimensional deformation in rolling by the finite-element method. Int. J. Mech. Sci. 26, 515–525. http://dx.doi.org/10.1016/0020-7403(84)90005-5.

Nalawade, R.S., Puranik, A.J., Balachandran, G., Mahadik, K.N., Balasubramanian, V., 2013. Simulation of hot rolling deformation at intermediate passes and its industrial validity. Int. J. Mech. Sci. 77, 8–16. http://dx.doi.org/10.1016/j.ijmecsci.2013.09.017.

Oh, S.I., Wu, W.T., Arimoto, K., 2001. Recent developments in process simulation for bulk forming processes. J. Mater. Process. Technol. 111, 2–9. http://dx.doi.org/10.1016/S0924-0136(01)00508-8.

Orowan, E., 1943. The calculation of roll pressure in hot and cold flat rolling. Proc. Inst. Mech. Engrs. 150, 140–167.

Pal, S.K., Talamantes-Silva, J., Linkens, D.A., Howard, I.C., 2007. Bond graph and finite element analyses of temperature distribution in a hot rolling process: a comparative study. J. Syst. Control Eng. 221, 653–661. http://dx.doi.org/10.1243/09596518JSCE281.

Phaniraj, M.P., Behera, B.B., Lahiri, A.K., 2005. Thermo-mechanical modeling of two phase rolling and microstructure evolution in the hot strip mill: Part I. Prediction of rolling loads and finish rolling temperature. J. Mater. Process. Technol. 170, 323–335. http://dx.doi.org/10.1016/j.jmatprotec.2005.05.009.

Polozine, A., Schaeffer, L., 2008. Influence of the inaccuracy of thermal contact conductance coefficient on the weighted-mean temperature calculated for a forged blank. J. Mater. Process. Technol. 195, 260–266. http://dx.doi.org/10.1016/j.jmatprotec.2007.05.020.

Priyadarshini, A., Pal, S.K., Samantaray, A.K., 2012. Finite element modeling of chip formation in orthogonal machining. In: Davim, J.P. (Ed.), Statistical and Computational Techniques in Manufacturing. Springer, Berlin, Heidelberg, pp. 101–144.

Rahman, B.M.A., Agrawal, A., 2013. Finite Element Modeling Methods for Photonics. Artech House, Boston, London.

Reddy, J.N., 2012. An Introduction to the Finite Element Method, third ed. Tata McGraw Hill Education Private Limited, New Delhi. http://dx.doi.org/10.2307/2007936.

Riahifar, R., Serajzadeh, S., 2007. Three-dimensional model for hot rolling of aluminum alloys. Mater. Des. 28, 2366–2372. http://dx.doi.org/10.1016/j.matdes.2006.08.011.

Rout, M., Pal, S.K., Singh, S.B., 2015. Cross rolling: a metal forming process. In: Davim, J.P. (Ed.), Mod. Manuf. Eng. Mater. Forming, Mach. Tribol. Springer International Publishing, Switzerland, pp. 41–64. http://dx.doi.org/10.1007/978-3-319-20152-8.

Rout, M., Pal, S.K., Singh, S.B., 2016a. Finite element simulation of a cross rolling process. J. Manuf. Process. 24, 283–292. http://dx.doi.org/10.1016/j.jmapro.2016.09.012.

Rout, M., Pal, S.K., Singh, S.B., 2016b. Finite element analysis of cross rolling on AISI 304 stainless steel: prediction of stress and strain fields. J. Inst. Eng. Ser. C. http://dx.doi.org/10.1007/s40032-016-0239-8.

Saad, Y., 2003. Iterative Methods for Sparse Linear Systems, second ed. Society for Industrial and Applied Mathematics, Philadelphia. 10.1016/S1570-579X(01)80025-2.

Said, A., Lenard, J.G., Ragab, A.R., Elkhier, M.A., 1999. Temperature, roll force and roll torque during hot bar rolling. J. Mater. Process. Technol. 88, 147–153. http://dx.doi.org/10.1016/S0924-0136(98)00391-4.

Sellars, C.M., 1985. Computer modelling of hot-working processes. Mater. Sci. Technol. 1, 325–332. http://dx.doi.org/10.1179/mst.1985.1.4.325.

Serajzadeh, S., Mahmoodkhani, Y., 2008. A combined upper bound and finite element model for prediction of velocity and temperature fields during hot rolling process. Int. J. Mech. Sci. 50, 1423–1431. http://dx.doi.org/10.1016/j.ijmecsci.2008.07.004.

Serajzadeh, S., Mirbagheri, H., Karimi Taheri, A., 2002. Modelling the temperature distribution and microstructural changes during hot rod rolling of a low carbon steel. J. Mater. Process. Technol. 125–126, 89–96. http://dx.doi.org/10.1016/S0924-0136(02)00322-9.

Shahani, A.R., Setayeshi, S., Nodamaie, S.A., Asadi, M.A., Rezaie, S., 2009. Prediction of influence parameters on the hot rolling process using finite element method and neural network. J. Mater. Process. Technol. 209, 1920–1935. http://dx.doi.org/10.1016/j.jmatprotec.2008.04.055.

Shangwu, X., Rodrigues, J.M.C., Martins, P.A.F., 1999. Simulation of plane strain rolling through a combined finite element-boundary element approach. Finite Elem. Anal. Des. 32, 221–233. http://dx.doi.org/10.1016/S0924-0136(99)00342-8.

Sheikh, H., 2009. Thermal analysis of hot strip rolling using finite element and upper bound methods. Appl. Math. Model. 33, 2187–2195. http://dx.doi.org/10.1016/j.apm.2008.05.022.

Shima, S., Mori, K., Oda, T., Osakada, K., 1980. Rigid-plastic finite element analysis of strip rolling. In: Proc. 4th Int. Conf. Prod. Engrs, pp. 82–87.

Sui, F.-l., Zuo, Y., Liu, X.-h., Chen, L.-q., 2013. Microstructure analysis on IN 718 alloy round rod by FEM in the hot continuous rolling process. Appl. Math. Model. 37, 8776–8784. http://dx.doi.org/10.1016/j.0061pm.2013.04.005.

Sun, C.G., 2005. Investigation of interfacial behaviors between the strip and roll in hot strip rolling by finite element method. Tribol. Int. 38, 413–422. http://dx.doi.org/10.1016/j.triboint.2004.09.002.

Sun, C.G., Hwang, S.M., 2000. Prediction of roll thermal profile in hot strip rolling by the finite element method. ISIJ Int. 40, 794–801. https://www.jstage.jst.go.jp/article/isijinternational1989/40/8/40_8_794/_pdf.

Synka, J., Kainz, A., 2003. A novel mixed Eulerian-Lagrangian finite-element method for steady-state hot rolling processes. Int. J. Mech. Sci. 45, 2043–2060. http://dx.doi.org/10.1016/j.ijmecsci.2003.12.008.

Tamano, T., 1973. Finite element analysis of steady flow in metal processing. J. Jpn. Soc. Tech. Plast. 14, 766.

Theocaris, P.S., 1985. A study of the contact zone and friction coefficient in hot-rolling. Metal Forming and Impact Mechanics. Pergamon Press Ltd., Oxford, pp. 61–90. http://dx.doi.org/10.1016/B978-0-08-031679-6.50012-6.

Thompson, E.G., 1982. Inclusion of elastic strain rate in the analysis of viscoplastic flow during rolling. Int. J. Mech. Sci. 24, 655–659.

Tieu, A.K., Jiang, Z.Y., Lu, C., 2002. A 3D finite element analysis of the hot rolling of strip with lubrication. J. Mater. Process. Technol. 125–126, 638–644. http://dx.doi.org/10.1016/S0924-0136(02)00371-0.

Vallellano, C., Cabanillas, P.A., García-Lomas, F.J., 2008. Analysis of deformations and stresses in flat rolling of wire. J. Mater. Process. Technol. 195, 63–71. http://dx.doi.org/10.1016/j.jmatprotec.2007.04.124.

Yang, Y.Y., Linkens, D.A., Talamantes-Silva, J., Howard, I.C., 2003. Roll force and torque prediction using neural network. ISIJ Int. 43, 1957–1966.

Yuan, S.Y., Zhang, L.W., Liao, S.L., Jiang, G.D., Yu, Y.S., Qi, M., 2009. Simulation of deformation and temperature in multi-pass continuous rolling by three-dimensional FEM. J. Mater. Process. Technol. 209, 2760–2766. http://dx.doi.org/10.1016/j.jmatprotec.2008.06.024.

Zhang, G.L., Zhang, S.H., Liu, J.S., Zhang, H.Q., Li, C.S., Mei, R.B., 2009. Initial guess of rigid plastic finite element method in hot strip rolling. J. Mater. Process. Technol. 209, 1816–1825. http://dx.doi.org/10.1016/j.jmatprotec.2008.04.038.

Zhang, S.H., Zhang, G.L., Liu, J.S., Li, C.S., Mei, R.B., 2010. A fast rigid-plastic finite element method for online application in strip rolling. Finite Elem. Anal. Des. 46, 1146–1154. http://dx.doi.org/10.1016/j.finel.2010.08.005.

Zienkiewicz, O.C., Jain, P.C., Onate, E., 1978. Flow of solids during forming and extrusion: some aspects of numerical solutions. Int. J. Solids Struct. 14, 15–38. http://dx.doi.org/10.1016/0020-7683(78)90062-8.

Numerical modeling methodologies for friction stir welding process

Rahul Jain, Surjya K. Pal, Shiv B. Singh
Indian Institute of Technology, Kharagpur, India

5.1 Introduction to FSW

Aluminum alloys have been on the radar of human race for its excellent strength-to-weight ratio and to achieve weight reduction without compromising the mechanical strength. However, weldability issues have limited the use of aluminum alloys for high strength applications such as aerospace, marine, etc. High thermal expansion coefficient, difference in solubility of gas (mainly hydrogen) in molten and solid states, and affinity for oxygen are the major bottlenecks in achieving high weld strength with fusion welding (Jain et al., 2015). Invention of friction stir welding (FSW) by The Welding Institute, Cambridge in 1991 (Thomas et al., 1991) eliminated the complexities in the welding of aluminum alloys. FSW is a remarkably simple process and is divided into three stages, viz., plunging, dwelling, and welding. In the first stage, a rotating, nonconsumable specially designed tool plunges onto the abutting edges of the clamped workpiece to generate heat at the interface (Fig. 5.1A). Interaction between the two materials generates heat and softens the workpiece. Afterward, the tool rotates for a while in its position to further increase the temperature of the workpiece in the dwelling stage (Fig. 5.1B). Lastly, the rotating tool travels along the abutting edges to perform welding through stirring of the material, as shown in Fig. 5.1C.

Note: Arrows show the tool movement in respective direction at various stages

The nonconsumable tool in FSW has two special features known as shoulder and pin. Shoulder-to-pin diameter ratio ranges from 3 to 5 depending upon the workpiece material and process parameters to achieve an efficient weld (Mishra and Ma, 2005). Design of the features is based on their functionality. Shoulder performs two vital functions: first, heat generation through friction and plastic deformation to soften the material. Second to contain the viscous material underneath it to avoid spilling of the material away from the welding zone, thus avoiding flash formation and thereby reducing loss of material. Shoulder diameter ranges from 12 to 20 mm, depending on the workpiece material, thickness, and process parameters (Arora et al., 2011a). The primary function of the pin is to stir the material and perform welding. It also contributes to the heat generation (Schmidt et al., 2004). FSW process holds several advantages over conventional fusion welding techniques such as lower distortion, higher

Computational Methods and Production Engineering. http://dx.doi.org/10.1016/B978-0-85709-481-0.00005-7

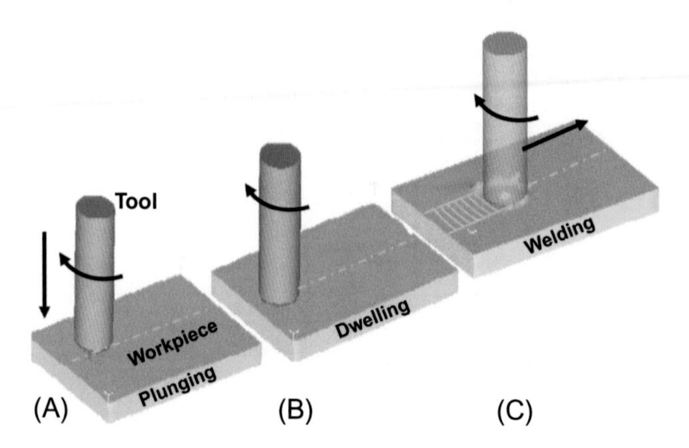

Fig. 5.1 Schematic representation of various stages of FSW: (A) plunging stage, (B) dwelling stage, and (C) welding stage. Note: *Arrows* show the tool movement in respective direction at various stages.

mechanical strength, finer grains, better energy efficiency (as compared to laser and electron beam welding), green technology, etc. (Jain et al., 2015). The only drawback is that the process leaves an exit hole at the end of welding.

5.1.1 Lexicon of FSW

Various nomenclatures used for FSW are illustrated in Fig. 5.2. Definition and significance of each terminology are as follows:

- **(i)** Tool shoulder: Part of the tool that comes in contact with the workpiece top surface and is mainly responsible for heat generation. It contributes 80%–90% of the total heat generation.
- **(ii)** Tool pin: Part of the tool that plunges inside the workpiece and contributes to the stirring of the material. It also contributes 10%–20% of the total heat generation.
- **(iii)** Advancing side (AS): Part of the workpiece where the tool rotation direction and the welding direction are same.
- **(iv)** Retreating side (RS): Part of the workpiece where the tool rotation direction and welding direction are opposite. Difference in relative velocity on either side of the tool makes FSW process asymmetric, with higher temperature and more deformation in AS as compared to RS.
- **(v)** Plunge depth (d): It is the distance by which the shoulder plunges onto the workpiece top surface. Plunge depth is provided to increase the contact area at the interface, for higher heat generation. Plunge depth generally ranges from 0.1 to 0.5 mm.
- **(vi)** Tilt angle (α): It is the angle by which the tool is tilted toward the trailing edge (or opposite to the welding direction). It is defined to increase the forging force on the workpiece to eliminate defects and also to reduce the flash formation. Tilt angle ranges from 0 degree to 3.5 degrees.

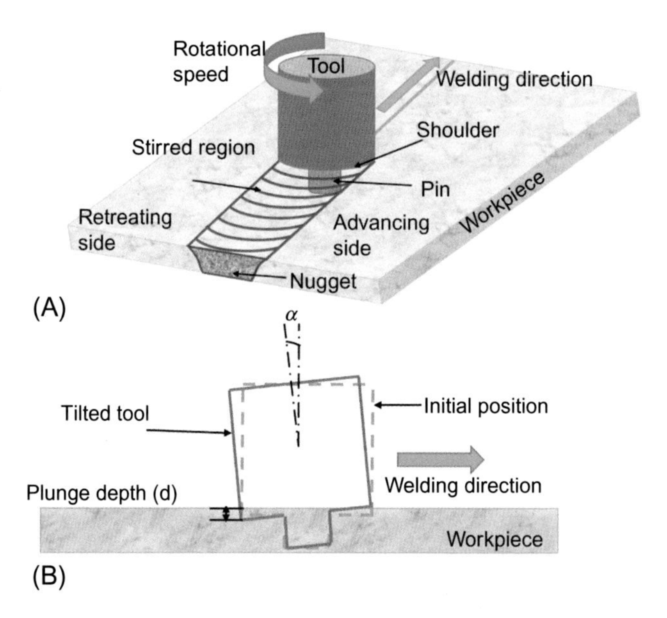

Fig. 5.2 Nomenclature of FSW: (A) schematic of isometric view and (B) section view.

5.2 Modeling of FSW: Requirement and complexities

Innovations invariably evolve through experiments in the laboratory and FSW is no exception. FSW was initially developed for aluminum alloys and later extended to other metallic alloys, dissimilar materials, polymers, etc. Computational methods should be tapped to support and guide experimental research particularly in terms of industrial application and minimization of development cost. Numerical models based on scientific understanding of the process and mechanism have great scope in optimizing the tool design, and process parameter window to achieve higher productivity and joint efficiency through the prediction of microstructure and defects. Numerical model is also handy for predicting parameters such as plastic strain, stress that influence microstructure and are difficult to be estimated experimentally.

FSW is a multiphysics problem where deformation of the material and temperature is interdependent. Displacement of the material generates heat due to plastic deformation and is reciprocated by reduction in yield strength at elevated temperature. This phenomenon should be captured through a coupled temperature-displacement model. Temperature during FSW is in the range of 0.7–0.9 times T_m (T_m is the melting point of the material in °C), which is above the recrystallization temperature of the material. This leads to evolution and formation of new grains, microstructure, and changes in the metallurgical/mechanical properties of the material. Contact condition between the tool and workpiece is nonlinear as friction and plastic deformation both contribute to heat generation. Experimental prediction of contact condition at elevated temperature is quite difficult. Sliding and sticking are two contact conditions possible at the interface and still it is not elucidated which condition resembles the actual situation. Former condition is generally used where the contact stress is lower than the yield strength of the material, and latter is defined for the process where contact stress is above the yield strength. Difference in velocity on the either side of the tool makes the process asymmetric, and this requires a complete modeling of workpiece contrary to the situation of fusion welding.

5.3 General steps for modeling a process

The fundamental steps of modeling are the same irrespective of the process. This section deals with the steps to be followed for the modeling of FSW irrespective of the chosen modeling method. Sequence might differ based on the commercial code chosen but steps remain the same. Following are the different aspects that need to be considered:

- complexity level and analysis type
- geometric modeling and assembly
- material property and constitutive equations
- simulation control and solver
- contact interaction
- boundary conditions
- mesh generation
- solution and postprocessing

Each of the steps is discussed in details in the subsequent sections. Elaboration is made based on FSW.

5.3.1 Complexity level and analysis type

At the inception of modeling, a researcher must identify the ultimate goal to select the appropriate level of complexity. Numerical and analytical methods have a role to play although the former has a dominance due to advanced workstations and commercially available software such as ABAQUS, ANSYS, DEFORM, etc. Numerical modeling works on the principle of discretizing the domain using finite element, finite volume techniques. These methods are capable of capturing complexity in geometry, boundary conditions, material constitutive behavior, but are computationally expensive if all the nonlinearities have to be incorporated. Therefore it is recommended to make assumptions for the simplification of the process that will not hamper the final results. A problem in three-dimensional can be simplified in two-dimensional, for example, in machining, making a orthogonal two-dimensional model (Priyadarshini et al., 2012), or plane strain condition for rolling operation, etc. Similarly, symmetry or axisymmetry of the process should be exploited to model quarter or half of the whole domain, for example, compression testing. It is also essential to validate the model with experiments and sensitivity of the model is known based on the assumptions, defined material properties, and boundary conditions. A two-dimensional FSW model will not be appropriate as output responses vary along the thickness direction, but the process can be assumed symmetric along the weld line. This is valid only if prediction of temperature distribution is the goal of the researcher although it had been reported that the values of temperature in AS and RS differ by 5–20°C (Jain et al., 2014). Different methods of analysis are discussed in Section 5.3.6.

5.3.2 Geometric modeling and assembly

Geometric models of the workpiece, tool, and backing plate are essential. Dimensions of the parts should be chosen based on the experiment performed. Geometric models could be created either in a CAD software and imported in the simulation environment, or created within the simulation environment. Once the parts are created, they should be assembled to resemble the initial position of the plunging stage, as shown in Fig. 5.3. Tool pin bottom is in contact with the workpiece top surface and backing plate is placed on the bottom face of the workpiece. Based on the analysis type, parts can be either rigid or deformable. In case of rigid part, strain and stress are not calculated. Calculation of temperature can be done by activating the thermal degree of freedom or defining heat transfer between the parts. In FSW, tool and backing plate can be defined as rigid bodies since their yield strength is much higher than the workpiece material. This assumption reduces the computation time.

Deformation of the workpiece can be defined based on two different theories discussed as follows.

a. *Elastic-plastic deformation*: In this method, both elastic and plastic deformations of the material are incorporated. Here, calculations of stress and strain at low strains are defined based on the linear elasticity principle, i.e., Hooke's law. The nonlinearity is associated with the plastic deformation. The subsequent calculations of stress with strain are related to the influence of strain hardening and are calculated based on the theory of plasticity by using a flow rule to correlate the increments of plastic strain with plastic flow (Laboratory, 2007). Based on the process, plastic deformation part can be time dependent or independent. In case of the former, the deformation type is known as elastic–viscoplastic, and for latter it is elastic–plastic. This method is fruitful for calculation of spring back effect and residual stress. This method requires small time increment for transient analysis, and it further reduces

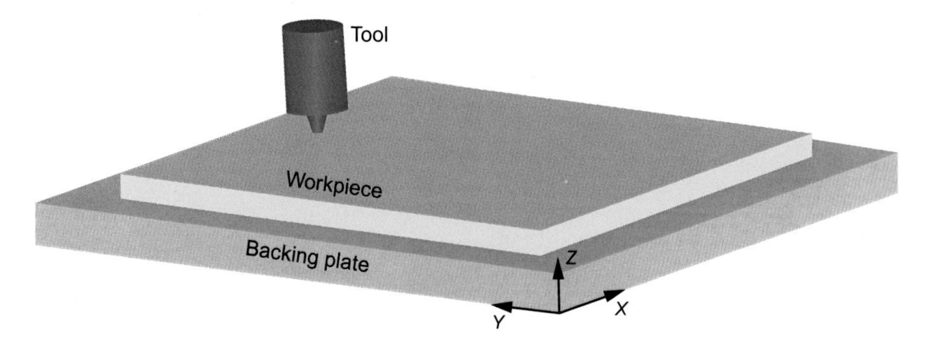

Fig. 5.3 Assembly of the tool, workpiece, and backing plate.

during the transition of deformation from elastic to plastic region (Kobayashi et al., 1989). This increases the computation time and sometimes leads to abnormal termination of the simulation due to nonconvergence. Although, this drawback could be tackled with mass scaling and by selecting good solver and with proper iterative method.

b. *Plastic deformation*: A simplified solution to the problems associated with elastic–plastic deformation is to neglect the elastic region and define the yield point as the inception of deformation. This assumption allows the choice of larger incremental step and reduces the chance of premature termination of the simulation. This assumption is valid for large deformation process like FSW where plastic strain is in a range of 6–80 (Nandan et al., 2008), as compared with the elastic deformation of 0.002 (based on offset method to calculate yield point). Spring back effect and residual stress cannot be calculated with this principle. Based on the process, plastic deformation can be classified as plastic and viscoplastic. In case of the former, flow behavior is independent of time, i.e., strain hardening is more prominent as compared to strain rate (e.g., cold forming process), while in the latter, strain rate is more dominant than strain, for example, hot working process. FSW is a hot working process as temperature is above the recrystallization temperature, and therefore flow behavior must be defined as strain rate sensitive.

5.3.3 Material properties and constitutive equation

Friction stir welding involves a large strain ranging from 6 to 80, moderate to high strain rate (5–100 s^{-1}) (Kuykendall et al., 2013), and is an elevated temperature deformation process. Temperature rises up to 0.9 times the absolute melting temperature of the material. Development of accurate numerical model for FSW requires a valid constitutive law throughout the process. A set of parametric equations, known as constitutive laws, are required to characterize the flow stress as a function of strain rate, strain, and temperature, which account for the prior strain, strain rate history of the material. Variety of constitutive laws have been reported in the literature, and among them Johnson-Cook model and Sheppard-Wright model are the most commonly used. Various material properties (thermal and mechanical) are to be defined based on the type of analysis, as presented in Table 5.1. Details of material models are discussed in the subsequent sections.

5.3.3.1 Jonson-Cook material model

The Johnson-Cook model (Johnson and Cook, 1983) is a multiplicative law. It is a strain, temperature-dependent viscoplastic model suited for high strain rate process, as expressed in Eq. (5.1).

$$\overline{\sigma} = [A + B\overline{\varepsilon}^{n}]\left[1 + C\ \ln\left(\frac{\dot{\overline{\varepsilon}}}{\dot{\varepsilon}_o}\right)\right]\left[1 - \left(\frac{T - T_{\text{room}}}{T_{\text{melt}} - T_{\text{room}}}\right)^{m}\right] \tag{5.1}$$

where, $\overline{\sigma}, \overline{\varepsilon}$, $\dot{\overline{\varepsilon}}$, $\dot{\varepsilon}_o$, T_{room}, and T_{melt} are flow stress, plastic strain, effective strain rate, reference strain rate (1 s^{-1}), room temperature, and melting temperature, respectively. A (in MPa) is the yield stress of the material at room temperature, B (in MPa) and n depict the influence of strain hardening and are known as hardening modulus and

Table 5.1 **Material properties to be defined in the modeling**

	Properties					
Analysis	**Young's modulus**	**Poisson's ratio**	**Density**	**Thermal conductivity**	**Heat capacity**	**Coefficient of thermal expansion**
Structural	✓	✓	✓	✖	✖	✖
Thermal	✖	✖	✓	✓	✓	✓
Coupled structural-thermal	✓	✓	✓	✓	✓	✓

work-hardening exponent, respectively. C and m are the strain rate hardening and thermal softening coefficient, respectively. Estimation of these material constants requires series of experiments under different conditions. Torsion test includes the effect of strain rate; therefore torsion and tensile tests at room temperature and at reference strain rate are used to calculate material constants A, B, and n. Similarly, constant C can be found out through series of torsion tests under different strain rates and plotting them between shear stress and shear strain. Thermal softening coefficient, m, is calculated through Hopkinson bar test at different temperatures (Johnson and Cook, 1983). Material constants of Johnson-Cook model for different material are presented in Table 5.2.

5.3.3.2 Sheppard and Wright model

Eq. (5.2) was initially developed by Sellars and Tegart (1972) and later modified by Sheppard and Jackson (1997) and Sheppard and Wright (1979), and is commonly known as Zener-Hollomon equation and is expressed in Eqs. (5.2) and (5.3).

$$\bar{\sigma}=\frac{1}{\alpha}\ln\sinh^{-1}\left[\left(\frac{Z}{A}\right)^{\frac{1}{n}}\right]=\frac{1}{\alpha}\ln\left\{\left(\frac{Z}{A}\right)^{\frac{1}{n}}+\left[1+\left(\frac{Z}{A}\right)^{\frac{2}{n}}\right]^{\frac{1}{2}}\right\} \tag{5.2}$$

$$Z=\dot{\bar{\varepsilon}}\exp\left(\frac{Q}{RT}\right)=A[\sinh(\alpha\bar{\sigma})]^{n} \tag{5.3}$$

where Z is the Zener-Hollomon parameter and Q (J/mol) is activation energy. A (s^{-1}), α (MPa^{-1}), Q, and n are material constants. R is the universal gas constant (J/mol/K). This equation is widely used specially for aluminum alloys because of availability of material constant (Sheppard and Jackson, 1997) and ease of implementation in the simulation environment. Detailed study on the various constitutive models and their influence on the FSW can be found in Kuykendall et al. (2013) and Kuykendall, 2011). Material constants for all the aluminum alloys can be found in (Sheppard and Jackson (1997), and some of them are mentioned in Table 5.3.

Table 5.2 Johnson-Cook material constant for different materials

	Material constant					
Material	***A* (MPa)**	***B* (MPa)**	***n***	***C***	***m***	**Analysis type**
AA 5059	167	596	0.551	0.001	1.0	CEL (Grujicic et al., 2012)
AA 2024-T3	369	684	0.73	0.0083	1.7	ALE (Schmidt and Hattel, 2004)
AA 7050	435.7	534.6	0.5	0.019	0.97	Lagrangian (Yu et al., 2012)

Table 5.3 Sheppard and Wright material model for different materials

Material	Material constant			
	ln A (s^{-1})	α (MPa^{-1})	n	Q (J/mol)
AA6061	19.3	0.045	3.55	145,000
AA2024	19.6	0.016	4.27	148,880
AA 7075	26.38	0.011	6.32	156,837

5.3.4 Contact interaction

Interactions in the numerical model have to be defined to emulate the physical phenomenon. Interactions among different parts are defined to enable heat and force transmissions. In FSW, two pairs interact mechanically and thermally. First pair is tool (shoulder and pin)–workpiece, where tool shoulder interacts with the top surface of the workpiece. Pin interacts with the workpiece. Second is the interaction between workpiece and backing plate, where bottom surface of the workpiece interacts with the top surface of the backing plate.

5.3.4.1 Mechanical interactions

Penalty contact algorithm is the most commonly used algorithm for nonlinear contact conditions. In this algorithm, penetration between two surfaces is restricted by a linear spring force or contact pressure which has a value proportional to the penetration depth, as shown in Fig 5.4. These forces at the contact tend to pull the surfaces into equilibrium position without any penetration (Grujicic et al., 2015). The contact condition can be defined as node-to-face penalty contact or face-to-face penalty contact.

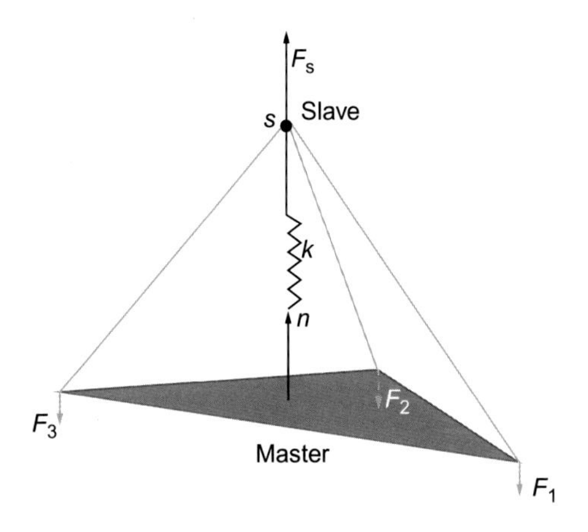

Fig. 5.4 Node-to-surface penalty contact element.

In case of the former method, each interacting surface pair consists of a dependent (slave) surface and an independent (master) surface. Slave surface is defined as nodes, whereas master surface consists of elemental faces. The elemental faces within one master surface should be such that any edge of any face has at most one neighboring face. Generally, mesh on the slave surface must be at least as fine as the master surface if not more (Penalty Contact Algorithm). In an interacting pair, if one part is the rigid body then it must be defined as a master surface and deformable part as a slave surface. If both the pairs are deformable then master and slave surfaces should be defined based on the yield strength of the material; the higher one should be defined as the master surface and lower as the slave surface. When the surfaces are in contact, contact pressure of any magnitude could be transmitted among them.

Tangential stress at the contact surface is defined through frictional condition. Friction between metal surface is due to two reasons: first, adhesion at the interface between the asperities of the contacting surfaces and second, plastic deformation at microscopic level due to relative velocity of the interacting pair (Mishra et al., 2014). Therefore frictional force is a function of load on the interacting surfaces and their physical properties. In FSW, deformation mainly takes place at the workpiece surface since tool is almost nondeformable in most of the cases. Coulomb's law of friction and shear friction law are two different models used in FSW. Coulomb's law of friction correlates normal force with the tangential contact force as expressed in Eq. (5.4).

$$\tau_s = \mu p \tag{5.4}$$

where τ_s, μ, and p are frictional shear stress, coefficient of friction, and normal compressive stress, respectively. It is a linear relation between the normal load and the tangential force. Latter increases linearly till a maximum value of yield strength of the material. Coefficient of friction has been defined as a constant value (Su et al., 2014), as a function of temperature (Biswas and Mandal, 2011), and in some cases as a function of slip rate and radius (Nandan et al., 2007). Some of the reported values of coefficient of friction are mentioned in Table 5.4.

Table 5.4 Various Coulombian frictional factor reported in literature for different materials

Material	Variation of μ	Coefficient of friction	Deformation	Analysis type
AA 6082	Constant	0.7	Viscoplastic	Lagrangian (Uyyuru and Kailas, 2006)
AA 6061	Constant	0.8	Elasto-viscoplastic	CEL (Al-Badour et al., 2013)
AA 2024	Constant	0.4	Viscoplastic	Eulerian (Tang and Shen, 2016)
Mild steel	$f(\omega, r, \delta)$	0.42–0.28	Viscoplastic	Eulerian (Nandan et al., 2007)

Shear friction model has been defined by a number researchers (Jain et al., 2014; Buffa et al., 2006a; Asadi et al., 2011) and is expressed as

$$\tau_s = mk \tag{5.5}$$

where m is shear factor which ranges between 0 and 1 and k is the shear strength of the material. Studies indicate that Eq. (5.5) is adequate for bulk forming process, while Eq. (5.4) is more appropriate for sheet metal forming process. This is because compressive normal stress in the sheet metal forming is much smaller in comparison with the bulk forming process (Kobayashi et al., 1989). Some of the reported shear factors for different materials are mentioned in Table 5.5. Since deformation in FSW is higher, therefore it also comes under the bulk forming process.

Combination of both the friction models can also be defined and is called as hybrid model. Here, Eq. (5.4) is followed till the deformation reaches shear strength of the material and Eq. (5.5) afterward. Selection of shear factor or coefficient of friction is very critical to the modeling as it influences the heat generation rate significantly. Smaller friction factor will lead to lower deformation and heat generation, accordingly it becomes difficult to stir the material. Higher factor will lead to abnormal heating and unreasonable temperature distribution.

Table 5.6 shows the comparison of simulation results of axial force and torque at different shear factors of 0.3, 0.4, and 0.5 with experimental values. Shear factor of 0.4 has predicted axial force and torque close to the experiment, as compared to other shear factor.

5.3.4.2 Thermal interaction

In FSW, frictional deformation and plastic deformation contribute to total heating. A percentage of total heat generated is transferred to the tool and backing plate through conduction and to ambient through convection and radiation. Remaining

Table 5.5 Various frictional factors reported in literature for different materials

Material	Shear factor	Analysis type	Deformation
AA 2024	0.4	Lagrangian (Jain et al., 2016a)	Viscoplastic
AA 7075	0.46	Lagrangian (Buffa et al., 2006a)	Viscoplastic
Alclad 7B04 aluminum alloy	0.32	Lagrangian (Zhao et al., 2015)	Viscoplastic
AZ 91 magnesium alloy	0.4	Lagrangian (Asadi et al., 2011)	Viscoplastic
Copper	0.4	Lagrangian (Pashazadeh et al., 2014)	Viscoplastic

Table 5.6 Influence of friction factor on axial force and torque (Jain et al., 2016a)

		Axial force (N)	Percentage error (%)	Spindle torque (N-m)	Percentage error (%)
Experimental		10,175	–	13.326	–
Simulation	$m=0.3$	11,353	11.5	10.65	20
	$m=0.4$	9500	6.6	11.93	10.5
	$m=0.5$	8186	19.6	17	27.6

amount is retained inside the workpiece to raise temperature. Fig. 5.5 shows the different modes of heat interaction in FSW process.

As stated earlier, total heat generated in FSW process is due to friction and plastic deformation, as expressed in Eq. (5.6). Frictional heating is incorporated through the tangential frictional stress defined at the interface (as discussed in Section 5.3.4.1) and it is equivalent to the product of frictional stress and slip rate at the interface as expressed in Eq. (5.7). Heat generation due to plastic deformation depends on the amount of deformation and the rate of heat generation is a product of flow stress and strain rate, as expressed in Eq. (5.8). Distribution of the temperature within the workpiece and the tool is governed by the Fourier law of heat conduction, as expressed in Eq. (5.9). The expressed Fourier law of heat conduction is for transient analysis, for steady-state analysis, the term on the right-hand side of the equation is equivalent to zero.

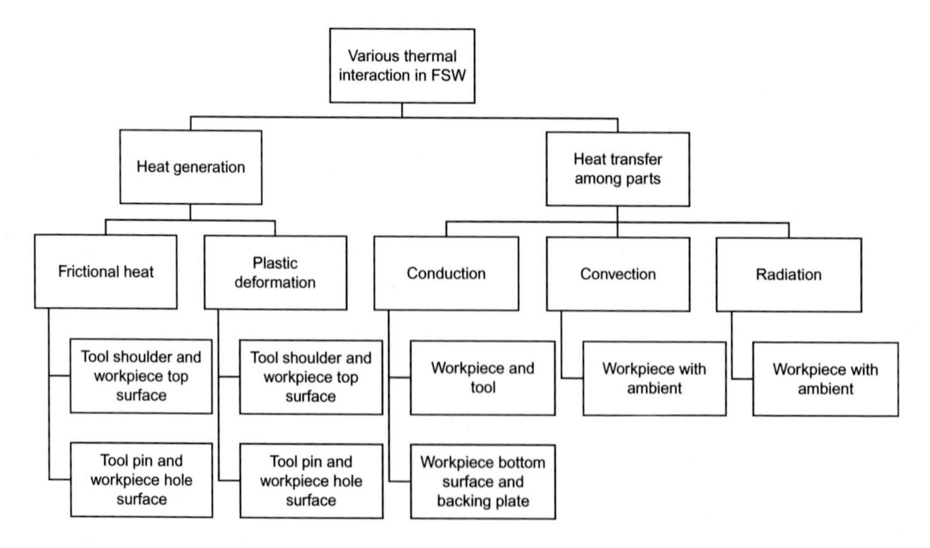

Fig. 5.5 Various interactions of heat transfer in FSW.

$$\dot{q} = \dot{q}_f + \dot{q}_p \tag{5.6}$$

$$\dot{q}_f = \phi(\tau_s \times \dot{\gamma}) \tag{5.7}$$

$$\dot{q}_p = \eta(\sigma \times \dot{\varepsilon}) \tag{5.8}$$

$$K\nabla^2 T + \dot{q} = \rho c \frac{\partial T}{\partial t} \tag{5.9}$$

where ρ, c, t, $\dot{q}$, $\dot{q}_f$, ϕ, $\dot{q}_p$, η $\dot{\gamma}$, and K are the mass density, specific heat capacity, time, heat generation rate, frictional heat generation rate, frictional heat factor, heat generation rate due to plastic deformation, inelastic heat fraction, slip rate, and thermal conductivity of the material, respectively. Frictional heat factor (ϕ) is the fraction of frictional energy converted to heat and this is reported as 1.0 in FSW (Asadi et al., 2011; Jain et al., 2016a; Buffa et al., 2006b). Inelastic heat fraction is the conversion factor for plastic deformation to heat energy. This ranges from 0.7 to 0.9 (Jain et al., 2016a; Nandan et al., 2006; Marzbanrad et al., 2014).

5.3.5 Boundary conditions

Workpiece is clamped with fixtures during experiment to eliminate its rigid rotation and translation. Also, tool is given rotational speeds and velocities in different direction to perform plunging and welding stages. Therefore boundary conditions must be defined during modeling to replicate the experimental conditions. Methods of defining the boundary condition are different in different techniques and therefore it would be discussed in detail in the relevant sections.

5.3.6 Mesh generation

In FEM the problem is approximated by converting the domain into finite number of distinct elements and solving it separately. This is achieved by dividing the domain into smaller units of simple geometry through a process called as discretization or meshing (Priyadarshini et al., 2012). Type and number of elements are chosen based on desired accuracy and available computing facility. Based on the modeling method, one-dimensional (line or beam), two-dimensional (triangular or quadrilateral), or three dimensional (tetrahedral or brick) elements can be chosen. Since FSW is a three-dimensional process, most appropriate element shape is either a quadrilateral or a brick element. Smaller element size increases the computational cost while coarser element hampers the accuracy of the result. Therefore a balance has to be struck between the two by performing mesh sensitivity analysis to find optimum element size such that variation in results is minimum and also computation time is acceptable. In FSW, mesh distortion is inevitable if Lagrangian formulation is used. This has to be tackled by defining a remeshing technique. Another option to avoid mesh distortion is using Eulerian technique in which instead of mesh nodes, only material points flow through the mesh. Details about both the methods and their implementation to simulate FSW are discussed in subsequent sections.

5.3.7 Simulation control and solver

Availability of commercial software has made FEM a more powerful tool for the researchers. Choosing an appropriate method for the modeling of FSW is of prime importance due to large deformation and nonlinearity involved. Different methods mainly used for the modeling of FSW are as follows:

- Lagrangian analysis
- Eulerian analysis
- Arbitrary Lagrangian and Eulerian (ALE) analysis
- coupled Eulerian-Lagrangian analysis (CEL)

Eulerian and Lagrangian analyses are the two classical methods for defining the movement in continuum mechanics. Both the methods differ in terms of movement of the nodes and material points. The other two methods, i.e., ALE and CEL are outcome of the first two methods where combinations of both the methods are used to overcome the individual drawbacks.

5.3.7.1 Lagrangian analysis

In this method, nodes and their associated materials points get displaced during the deformation of the domain. Throughout the problem nodes on the boundary surfaces remain along the element edge; this eventually makes the definition of boundary and contact conditions easier. Quadrature points (points at which results are calculated) coincide with the material points therefore each time constitutive equations are evaluated at the same material points leading to straightforward calculation of constitutive equation. This makes Lagrangian analysis best suited for solid mechanics or process with lower deformation. In case of FSW, this approach will lead to large mesh distortion and may lead to nonconvergence of the solution or premature failure of the simulation. This issue can be tackled by coupling Lagrangian method with a remeshing technique, such that every time mesh distorts beyond a critical limit, it is remeshed back to original shape. Details on the remeshing is discussed in Section 5.4.2.2.

Lagrangian method can be categorized into two types, viz., total Lagrangian and updated Lagrangian. In case of the former method, integral of weak form is performed over the initial configuration and derivatives are calculated with respect to the material coordinates. In case of the latter, integral of the weak form is carried out over the current or deformed configuration and the derivatives are calculated with respect to the spatial coordinates. A lot of researchers (Jain et al., 2014; Buffa et al., 2006a; Kheireddine et al., 2013; Trimble et al., 2012) have used Lagrangian method to simulate FSW process due to easy tracking of the material and boundaries. At the same time, Lagrangian method can model the material flow in FSW with particle tracking method. A detailed methodology for the simulation of FSW by using Lagrangian method is discussed in Section 5.4.

5.3.7.2 Eulerian analysis

Contrary to the Lagrangian analysis, in Eulerian method, mesh nodes are fixed in space and material points are allowed to flow through it. Since mesh is fixed in space, problem associated with mesh distortion is eliminated. Major drawback of this method is the difficulty to model free boundary surface as boundary nodes may not be coinciding with the element nodes and therefore it can only be used when the boundaries of the deformed surface are known a priori (Priyadarshini et al., 2012). Therefore plunging stage of FSW cannot be modeled with Eulerian analysis, and thus the impression of the tool shape has to be created before initiating the simulation. Methodology for the simulation of FSW with Eulerian analysis is discussed in Section 5.5.

5.3.7.3 Arbitrary Lagrangian-Eulerian analysis

Eulerian and Lagrangian methods have their own pros and cons. In ALE method, the nodes can move arbitrarily such that advantages of Lagrangian and Eulerian methods are exploited to minimize their drawbacks. Freedom of moving mesh allows larger distortions of continuum that can be handled as compared to Lagrangian method. However, mesh movements have its limitation and very large deformation is difficult to handle. Nevertheless, attempts have been made to model FSW with this method but they are limited (Zhang et al., 2007; Zhang and Zhang, 2009; Soundararajan et al., 2005). For detailed theory of ALE methods, readers may refer the work of Donea et al. (1999).

5.3.7.4 Coupled Eulerian-Lagrangian analysis

CEL method is also an attempt to capture the strength of both Eulerian and Lagrangian methods. This method is used for application in the area of fluid structure interaction analysis (FSI) involving large deformation. In this method, a Lagrangian frame is defined to discretize the moving structure or solid structure and Eulerian frame is defined to discretize the fluid domain. For the representation of different domains, boundary of Lagrangian domain is taken as reference. Contact interface between Lagrangian and Eulerian parts is defined by the velocity of Lagrangian boundary as a kinematic constraint in the Eulerian calculation and the stress from the Eulerian domain is used to calculate the resulting surface stress on the Lagrangian domain (Benson, 1992).

5.3.8 Capabilities of software

Various available commercial codes make use of different techniques as discussed earlier and are suitable for modeling FSW. Researchers have used different software such as DEFORM (Jain et al., 2016a,b), ABAQUS (Schmidt and Hattel, 2004; Zhang and Zhang, 2009), ANSYS-Fluent (Su et al., 2014), Forge (Assidi et al., 2010), etc., to simulate FSW. Comparison among a few commonly used software is shown in Table 5.7. Other methods used for simulation are smooth particle hydrodynamics (SPH) (Das and Cleary, 2016), natural element method (Alfaro et al., 2009), etc. In the subsequent section, modeling of FSW with Lagrangian, Eulerian, and CEL method is discussed in details.

Table 5.7 Comparative analysis of capabilities of different methods

Analysis	Software	Deformation type	Temperature analysis	Phase of FSW	Strain and strain rate	Material flow	Microstructure	Time	Defects
Lagrangian	DEFORM-3D	Visco-plastic	Transient	All phase	Yes	Point tracking (Jain et al., 2016a)	Yes (Asadi et al., 2015)	Moderate	—
	Forge 3D	Visco-plastic	Transient	All phases	Yes	Not reported	Not reported	High	—
ALE	ABAQUS/ Explicit	Elastic-visco-plastic	Transient	Plunging neglected	Yes	Point tracking	With subroutine	High[a]	—
Eulerian	Fluent or Star CCM+	Visco-plastic	Steady or transient	Plunging neglected	Yes	Stream line	Not reported	Low	—
CEL	ABAQUS/ Explicit	Elastic-visco-plastic	Transient	All phases	Yes	Marker material (Grujicic et al., 2012)	Not reported	High[a]	Yes

[a] Computational time can be reduced by using mass scaling.

5.4 Modeling of FSW with Lagrangian analysis

Current state-of-the-art research work in application of updated Lagrangian method for modeling FSW is discussed. DEFORM-3D has been used to simulate the process because of its strong in-built remeshing technique and material library. One of the initial works reported in the area of FSW modeling based on ALE method is by Xu et al. (2001). They developed a two-dimensional steady-state model with plane strain condition in ABAQUS to simulate the material flow in FSW. A group of researchers has developed a three-dimensional thermomechanical model based on updated Lagrangian method and extended that model for pin design (Buffa et al., 2006b), FSW T-shape (Fratini et al., 2010), dissimilar material (Buffa et al., 2015), FSW of steel (Buffa and Fratini, 2009). Researchers have also applied the Lagrangian model to study material flow (Tutunchilar et al., 2012), microstructure (Asadi et al., 2015), and pin shape (Jain et al., 2016a; Jain et al., 2017).

5.4.1 Geometric modeling and material model

Workpiece is defined as rigid viscoplastic material where elastic deformation is neglected and inception of deformation is considered from the yield point. This assumption is reasonable since FSW is a large deformation process and it makes convergence of the solution faster. The workpiece is modeled as a single continuum body instead of two separate sheets to avoid nonconvergence arising due to contact at the faying surfaces of the workpiece (Jain et al., 2016a). Workpiece thickness in the present work is taken as 3.1 mm with a total length and width of 110 and 114 mm, respectively. Cylindrical tool was considered in the analysis and was modeled as a rigid body. This is a valid assumption as yield strength of the tool material is much higher than the workpiece. Tool shoulder is defined as 16 mm, having a cylindrical pin with diameter and height of 5 and 2.5 mm, respectively. Backing plate is also considered as a rigid body with a length and width of 150 mm and 150 mm, respectively, and its thickness is defined as 12 mm. The tool is tilted by 2 degrees toward the negative Y direction, as shown in Fig. 5.6. AA 6061-T6 is defined as the workpiece material and Sheppard and Wright model as expressed in Eq. (5.3) is used to define the flow stress behavior for different temperature and strain rate. Material constants are defined in Table 5.8. Tool steel H13 is taken as the tool and backing plate materials. The workpiece and the tool material and their dimensions were consistent with the experimental work.

For coupled analysis various mechanical and physical properties of the material are defined as mentioned in Table 5.9.

5.4.2 Mesh generation

In this analysis coupled tetrahedral elements have been defined to discretize the workpiece. Each node of the element has two degrees of freedom, i.e., temperature and displacement. Since the tool and backing plate are defined as a rigid body, therefore

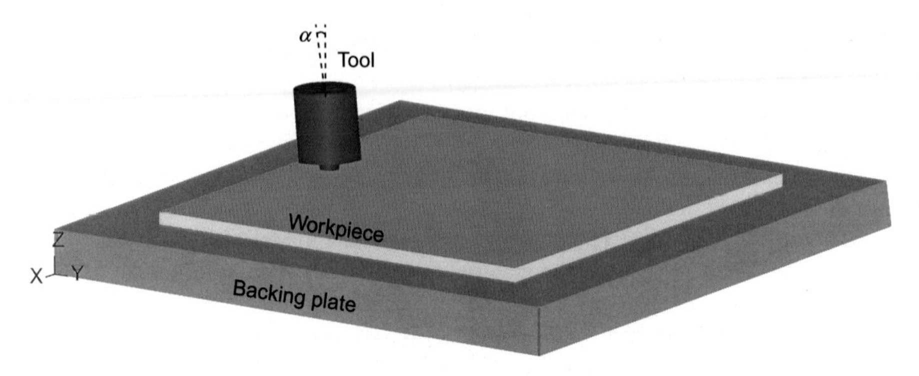

Fig. 5.6 The initial position of the parts in the assembly.

Table 5.8 Material constant for AA 6061-T6 for Sheppard and Wright model (Sheppard and Jackson, 1997)

A (s^{-1})	α (MPa^{-1})	Q (J/mol)	n	R (J/mol/°C)
240×10^6	0.045	145,000	3.55	8.314

Table 5.9 Physical properties of workpiece, tool, and backing plate (DEFORM v 11.0 documentation, 2015)

Properties	AA 6061	Tool steel H13
Young's modulus (N/mm^2)	68,950	210,000
Thermal conductivity (N/s°C)	181	24.5
Heat capacity (N/mm^2°C)	2.43	2.78
Coefficient of thermal expansion (μmm/mm°C)	22	11.7
Poisson's ratio	0.33	0.3
Emissivity	0.7	0.7

they are meshed with tetrahedral elements with single degree of freedom, which is temperature. Meshed assembly of the model is shown in Fig. 5.7.

Finer mesh is desired at the interaction zone of the tool and workpiece to capture the output responses accurately. To achieve this a mesh density window is defined with a constant mesh size of 0.6 mm (shown in Fig. 5.7). Mesh size of 0.6 mm is selected based on mesh refinement analysis as discussed subsequently. Remaining domain of the workpiece is meshed with a minimum element size of 2.25 with a size ratio of 2. Workpiece is meshed with 74,027 elements. Tool is meshed with 18,343 elements with a minimum element size of 0.5 mm and a size ratio of 2. Backing plate is meshed with 2646 elements with a minimum element size of 6 mm and size ratio of 2.

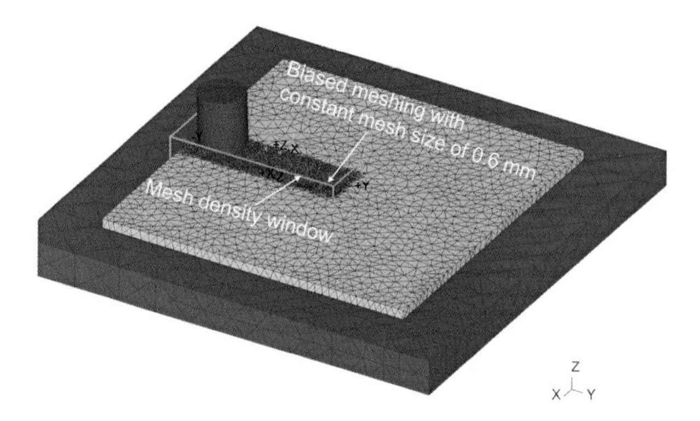

Fig. 5.7 Meshed assembly of the models.

5.4.2.1 Mesh refinement study

Apart from element type, size of an element plays a vital role in determining the accuracy and computational time. Finer mesh leads to higher computation time while coarser produces inaccurate results. Therefore to strike a balance between the accuracy and computational time, mesh sensitivity study must be carried out where mesh size is changed from coarser to finer for the same set of parameters and their output responses are compared. Five different sizes of mesh as shown in Fig. 5.8 have been considered and various output responses are compared.

Table 5.10 presents the influence of mesh size on computational time and output responses such as axial force and spindle torque. Maximum and minimum computational time are recorded for mesh size of 0.5 and 1 mm as 106 h and 17 h 41 min, respectively, on 2.9 GHz 4 core Intel i5 processor with 8 GB RAM. As element size reduces computational time increases due to increase in the total number of elements which in turn increases the size of the element stiffness matrix. Spindle torque is mostly unaffected with the mesh size as the deviation between maximum and minimum value is 2.5% but mesh size has an influence on axial force. Finer mesh size leads to increase in force and it is also observed that difference between axial forces for the

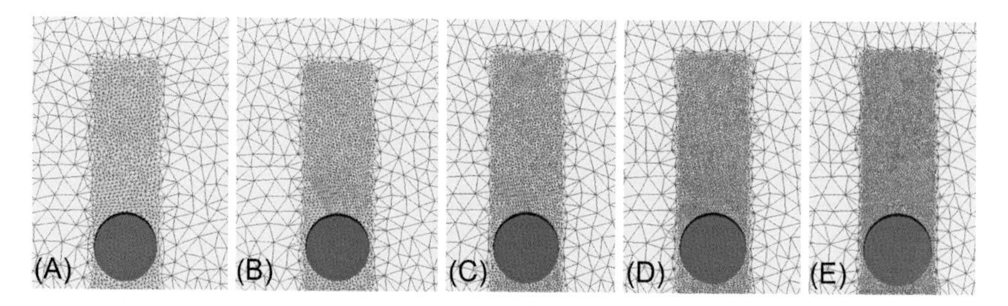

Fig. 5.8 Mesh size for mesh refinement analysis: (A) 1 mm, (B) 0.8 mm, (C) 0.7 mm, (D) 0.6 mm, and (E) 0.5 mm.

Table 5.10 Influence of mesh size on computational time, axial force, and spindle torque

	Mesh size (mm)				
Output response	**1**	**0.8**	**0.7**	**0.6**	**0.5**
Total elements	42,912	54,603	55,779	74,027	100,223
Computational time (h:min)	17:41	19:37	23:47	37:00	106:00
Torque (N-m)	16.3	16.06	15.95	15.94	15.9
Axial force (N)	6502	7749	8701	9276	9594

mesh sizes of 0.6 and 0.5 mm is only 3.4% indicating that axial force does not change with further mesh refinement. Based on Table 5.10, mesh size of 0.6 mm seems to be the optimum mesh size, but other output responses such as temperature, material velocity, and strain rate should also be analyzed before drawing conclusion. Gradient of strain rate, material velocity, and temperature exist along the transverse direction of the weld, therefore mesh refinement study is carried out at five predefined points as shown in Fig. 5.9.

Points are equidistant with a distance of 2 mm and are ahead of the pin center by 10 mm. Two points (P1, P2) are on the AS, and P4 and P5 are on the RS. Three points (P2, P3, and P4) are chosen in the stirred zone and remaining ones are in the heat-affected zone. Results for different mesh sizes taken based on the points, shown in Fig. 5.9, are shown in Fig 5.10. Fig. 5.10A shows the temperature for different mesh size for points P1–P5. Temperature remains almost constant with the change in the mesh size. Though mesh size of 1 mm is showing lower temperature for P3 and P4, the deviation is only 4% for both the cases. Fig. 5.10B shows the mesh refinement study on average material velocity. P1 and P5 are away from the stirred zone and hence material velocities are close to zero and are almost same for all sizes of the mesh. Material velocity is also same for point P4, which is on the RS. Mesh sizes of 1, 0.8, and 0.7 mm predict lower velocity as compared with mesh of 0.6 and 0.5 mm for points P2 and P3. Deviation in material velocity for the latter is 1.6%.

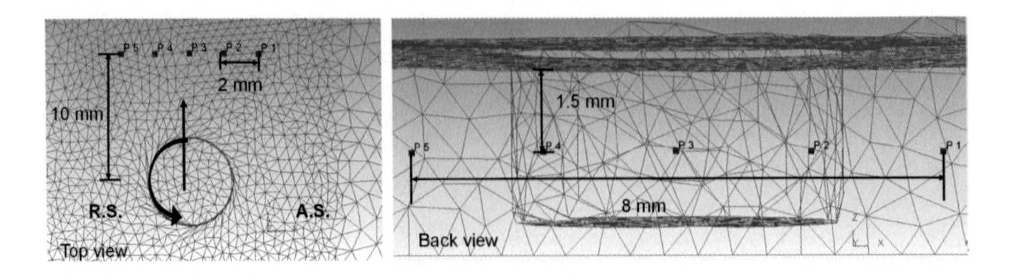

Fig. 5.9 Selected points for mesh refinement study.

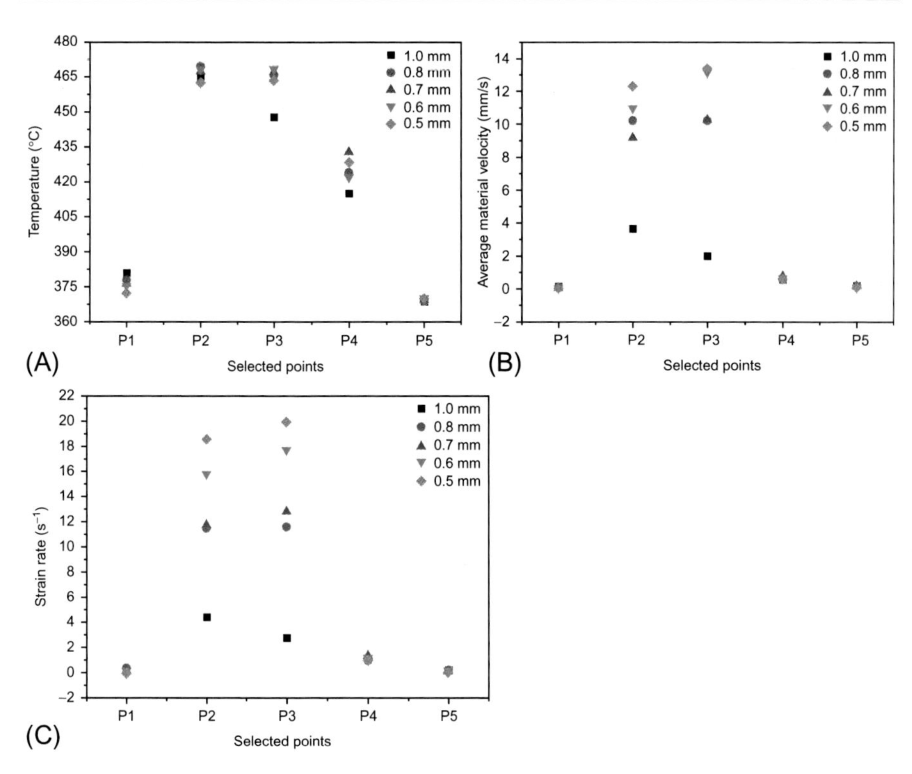

Fig. 5.10 Results of mesh refinement study: (A) temperature, (B) average material velocity, and (C) strain rate.

Fig. 5.10C shows the mesh refinement study for strain rate. Strain rate has a steepest gradient along the transverse direction and exists for a shorter distance. Therefore maximum strain rates observed for points P1 and P5 are 0.34 s^{-1} and 0.24 s^{-1}, respectively, and are almost the same for all selected mesh sizes. But in the stirred zone, variation in the strain rate is evident. Coarser meshes (1 mm, 0.8 mm, and 0.7 mm) have predicted much lower strain rates as compared to 0.6 and 0.5 mm. The deviations between 0.6 mm and 0.5 mm mesh size are 12% and 17% for P3 and P2, respectively. The overall mesh refinement study indicates that output variables have different behaviors for different mesh sizes and user should select mesh size based on the desired output responses. Torque and temperature are mostly independent of the studied mesh size and researchers can select element size of 1 mm or 0.8 mm. Along with torque and temperature, if a researcher is interested in studying material velocity and axial force, then 0.6 mm mesh size is an optimum value, in the adopted model. Here, computation time is not on the higher side and also results do not deviate very much even if a finer mesh is chosen. Though mesh size of 0.6 mm also provides adequate accuracy for strain rate, researchers should go for finer mesh and higher computational configuration for the same as the deviation in case of strain rate is more than 10%.

5.4.2.2 Remeshing technique

It is mandatory to define a remeshing technique with Lagrangian analysis, else the solution will abort due to excessive deformation of the mesh. Also, large deformation can shrink or stretch elements into unacceptable shapes leading to nonconvergence of the solution because of negative Jacobian. Remeshing procedure consists of two steps: first, assignment of new mesh on the domain and this is achieved with the conventional mesh algorithm used for meshing the domain and second, transfer of variables from old mesh to new mesh. Temperature and effective strains are two variables that depend on the deformation history, and hence it is interpolated to the new mesh. Temperature is calculated at the nodal points, and thus it is expressed by the element shape function over the domain and interpolation from old to the new mesh is done by calculating it at the new node locations. Effective strain interpolation on the new mesh requires additional step, as it is calculated at the quadrature points. Therefore prior to interpolation, effective strain at nodal points is estimated from the values given at the quadrature points. Afterward, its distribution is expressed by nodal point values and element shape functions. Area-weighted average scheme is used to interpolate the results from old to new mesh.

5.4.3 Contact interaction

Contact between the tool and workpiece is defined based on shear friction law as defined in Eq. (5.5) with a constant shear factor of 0.4. Contact between the workpiece and backing plate is defined based on coulomb's law of friction with a coefficient of friction of 0.2. In both the cases, workpiece is defined as the slave object; tool and backing plates are defined as the master surface. Various heat transfer phenomena have been defined to incorporate the heat loss to the tool, backing plate, and to environment, as shown in Fig. 5.11.

Convective heat transfer coefficients between the interacting pair of tool–workpiece and workpiece–backing plate are defined such that total convective heat transfer between them is equivalent to the conduction in the experiment as expressed in Eqs. (5.10) and (5.11), respectively. Eq. (5.12) incorporates the convective and radiation heat transfers between all surfaces of the workpiece (except bottom face) and environment.

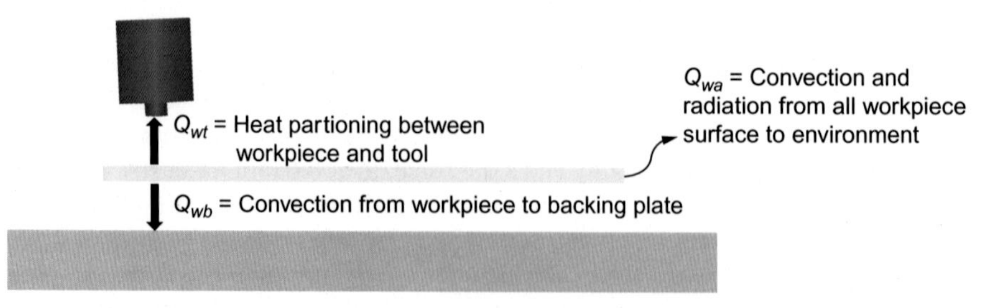

Fig. 5.11 Thermal interactions between different parts and environment.

$$Q_{wt} = K_{wt}\frac{\partial T}{\partial z} = h_t(T_w - T_t) \tag{5.10}$$

$$Q_{wb} = K_{wb}\frac{\partial T}{\partial z} = h_b(T_w - T_b) \tag{5.11}$$

$$Q_{wa} = \sigma_b \varepsilon_b \left(T_w^4 - T_a^4\right) + h_a(T_w - T_a) \quad (n \text{ is direction vector}) \tag{5.12}$$

where Q_{wt}, K_{wt}, h_t, T_w, T_t, Q_{wb}, K_{wb}, h_b, T_b, Q_{wa}, σ_b, ε_b, h_a, and T_a are heat transfer from workpiece to the tool, conductance between the workpiece and tool, convective heat transfer coefficient between the workpiece and the tool, workpiece temperature, tool temperature, heat transfer between workpiece and backing plate, conductance between workpiece and backing plate, convective heat transfer between the workpiece and backing plate, backing plate temperature, total heat transfer between workpiece and environment, Stefan-Boltzmann constant, emissivity of the workpiece, convective heat transfer between the workpiece and environment, and environment temperature, respectively. Values of convective heat transfer coefficient and other constants are defined in Table 5.11.

5.4.4 Boundary conditions and assumptions for the model

Various boundary conditions are defined on the workpiece to replicate the clamping condition during experiment. Side faces of the workpiece are constrained in X, Y, and Z directions to arrest any rigid motion. Rotational and translational movements are defined on the tool to consider different phases of FSW as shown in Fig. 5.1. Tool is rotated about Z axis and travel velocity is defined in Z and Y axes for plunging and welding phase, respectively. Following are the assumptions made during modeling: (a) physical properties of the workpiece and tool are defined as independent of temperature, (b) uniform heat transfer between the bottom face of the workpiece and backing plate is considered, (c) shear factor is defined as a constant value, i.e., it is independent of temperature, and (d) tool and backing plate are defined as a rigid body.

Table 5.11 **Parameters for heat transfer coefficients**

Variables	Values
Heat transfer between workpiece-tool and environment (h_a) (N/mm/s/°C)	0.02
Heat transfer between workpiece and tool (h_t) (N/mm/s/°C)	11
Heat transfer between workpiece bottom surface and backing plate (h_b) (N/mm/s/°C)	3
Stefan-Boltzmann constant (N/mm/K^4)	5.669×10^{-11}

5.4.5 Governing equation

The governing equation of the numerical model describes about the methodology for the calculation and change of the dependent (unknown) variables. The proposed method is an implicit method as the governing equation has velocity as an unknown variable for the calculation of change of dependent variables with respect to time. Governing equation for this method consists of three equilibrium equations, as expressed in Eq. (5.13) in notation form, yield condition expressed in Eq. (5.14), and five strain rate ratios derived from the flow rule.

$$\text{Equilibrium condition}: \frac{\partial \sigma_{ij}}{\partial x_j} = 0 \quad (i, j = 1,2,3) \tag{5.13}$$

$$\text{Yied critertion}: f(\sigma_{ij}) = C, \quad \overline{\sigma} = \sqrt{\frac{3}{2}\left(\sigma'_{ij}\sigma'_{ij}\right)} \tag{5.14}$$

$$\text{Constitutive equation}: \dot{\varepsilon}_{ij} = \frac{3\dot{\overline{\varepsilon}}}{2\overline{\sigma}}\sigma'_{ij} \quad \text{with} \quad \dot{\overline{\varepsilon}} = \sqrt{\frac{2}{3}\{\dot{\varepsilon}_{ij}\dot{\varepsilon}_{ij}\}} \tag{5.15}$$

$$\text{Combatibility conditions}: \dot{\varepsilon}_{ij} = \frac{1}{2}\left(\frac{\partial u_i}{\partial x_j} + \frac{\partial u_j}{\partial x_i}\right) \tag{5.16}$$

$$\text{Flow rule}: \dot{\varepsilon}^{pl} = \dot{\beta}\frac{\partial f(\sigma_{ij})}{\partial \sigma_{ij}} \tag{5.17}$$

where σ, σ'_{ij}, $\dot{\varepsilon}_{ij}$, u_i, $\dot{\beta}$, and f are Cauchy stress, deviatoric stress, strain rate component, velocity component, proportionality constant, and yield function, respectively. The boundary conditions are defined in terms of traction and velocity. Frictional contact condition is used to define traction as discussed in Section 5.4.3. Velocity boundary conditions are discussed in Section 5.4.4. Achievement of complete solution that satisfies all the governing equations is difficult and therefore it is solved by using a variational principle used in FEM. It is based on the one of the two variational principles. Here, among the admissible velocities u_i fulfill the conditions of compatibility, incompressibility, and velocity boundary conditions. The actual solution gives the following functional (function of a function) a stationary value (Rout et al., 2016),

$$\pi = \int_V E(\dot{\varepsilon}_{ij})dV - \int_{S_F} F_i u_i dS \tag{5.18}$$

where F_i, V, S_F, and E_{ij} are surface traction, volume of the workpiece, force surface, and work function, respectively. Based on second extremum principle for rigid viscoplastic material, work function is given by $\sigma'_{ij} = \frac{\partial E}{\partial \dot{\varepsilon}_{ij}}$. The solution to the original boundary value problem is obtained from the solution of the dual variational problem, where the first-order variation of the functional vanishes, as given in Eq. (5.19).

$$\delta\pi = \int_V \bar{\sigma}\delta\dot{\bar{\varepsilon}}dV - \int_{S_F} F_i\delta u_i dS = 0 \tag{5.19}$$

Incompressibility in penalized form is utilized to remove the incompressibility constraint on admissible velocity field. With this, Eq. (5.20) forms the fundamental equation for finite element formulation,

$$\delta\pi = \int_V \bar{\sigma}\delta\dot{\bar{\varepsilon}}dV + \lambda\int_V \dot{\varepsilon}_v\delta\dot{\varepsilon}_v dV - \int_{S_F} F_i\delta u_i dS = 0 \tag{5.20}$$

where λ and δu_i are large penalty constant and arbitrary variation, respectively. $\delta\dot{\bar{\varepsilon}}$. and $\delta\dot{\varepsilon}_v$ are variations in strain rate derived from δu_i. $\dot{\varepsilon}_v = \dot{\varepsilon}_{ii}$ is the volumetric strain rate. For detailed formulation of governing equation, readers may follow reference (Kobayashi et al., 1989).

5.4.6 Solvers and iterative method

DEFROM-3D has two different solvers, viz., sparse solver and conjugate gradient (CG) solver. Sparse provides direct solution by using the sparseness of the FEM formulation to improve speed. CG tries to solve the problem iteratively by approximating the solution. It has several advantages over sparse such as lower solving time and memory, but in some cases such as lower contact among dies, rigid motion of deformable object, etc., sparse solver converges but CG might not converge. Therefore in this analysis mainly CG solver is used but at times when the solution is not converging it switches to sparse solver for convergence and then again switches back to CG. Direct iteration (DI) and Newton–Raphson (NR) are two iterative methods available. In DI, it is assumed that the constitutive equation is linear during each iteration. The procedure for DI method is shown in Fig. 5.12 (Kobayashi et al., 1989):

5.4.7 Results and discussion

5.4.7.1 Temperature and plastic strain distribution

Contours of temperature and plastic strain are shown in Fig. 5.13A and B, respectively, for rotational speed of 900 rpm and a welding speed of 60 mm/min. Maximum temperature of 455°C is observed in the stirring zone just below the shoulder and is 0.78 times the melting point (in °C) of the material. The spread of the temperature contour ahead of tool is lower as compared to the trailing side. This is because colder material comes in contact with the tool and during stirring gets heated up and is deposited at the trailing edge of the tool leading to higher and more uniform temperature. Section A-A of Fig. 5.13A shows the temperature distribution in the transverse

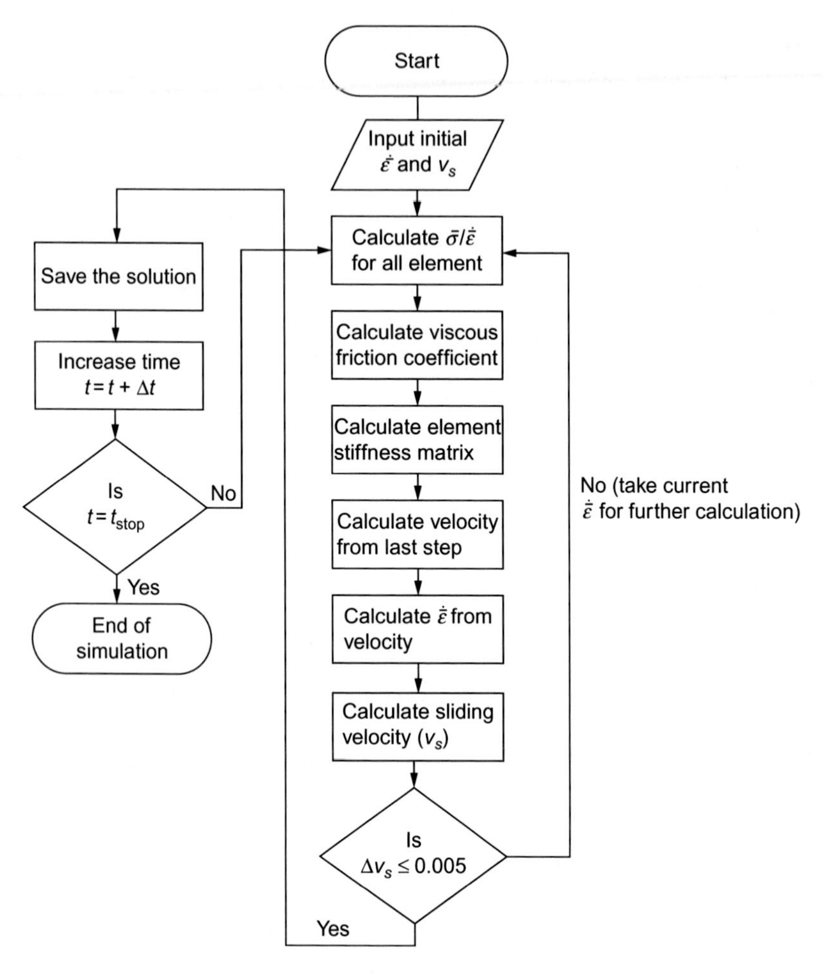

Fig. 5.12 Flowchart for direct integration iterative method.

direction. Temperature distribution along the thickness direction has typical V shape because of the presence of higher heat source in form of shoulder on the top surface and also due to much higher heat dissipation from bottom face of the workpiece to the backing plate as compared to the top surface. Fig. 5.13B shows the plastic strain distribution on the top surface and along the transverse direction. Higher plastic strain is observed at the AS, as compared to the RS due to difference in relative velocity. Asymmetry in plastic strain distribution is higher as compared to the temperature and is due to the mechanism of material flow during FSW. Material on the AS is stirred through the RS before getting deposited on the trailing edge of the tool, and on the contrary material on the RS is just displaced before getting deposited. This leads to higher displacement for the material on the AS as compared to that on the RS, and hence higher strain is achieved on the AS.

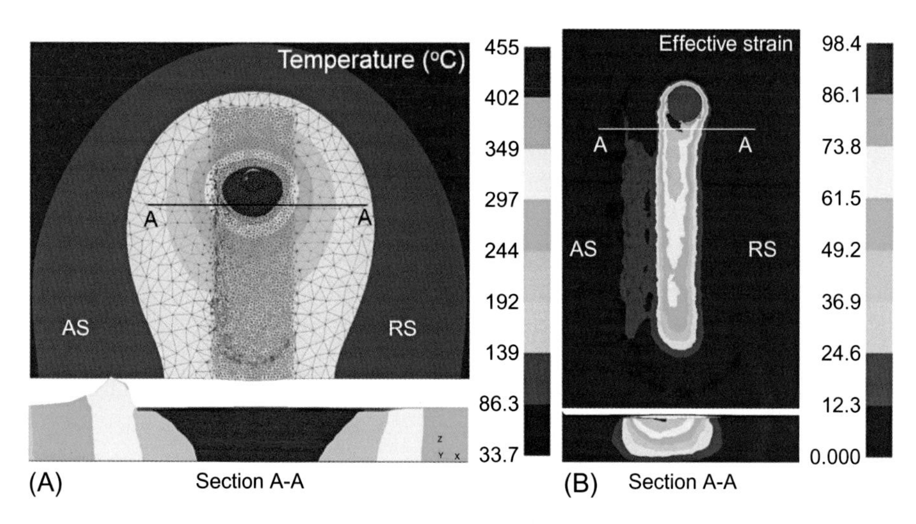

Fig. 5.13 (A) Temperature contour on top surface and traverse section and (B) plastic strain contour on top surface and traverse section.

Experimentally measured temperature is used to validate the simulated result as shown in Fig. 5.14A for rotational speed of 900 rpm and 60 mm/min. K-type thermocouples were embedded in the workpiece at a depth of 1.5 mm for measuring the inside temperature. In X and Y directions the thermocouple was located at a distance of 8 and 33 mm, respectively, according to the coordinate system shown in Fig. 5.6. Temperature predicted by the model is in good agreement with the experimental result. The difference between the model-predicted maximum temperature and the corresponding measured temperature is merely 5°C. Fig. 5.14B shows the temperature variation in the transverse direction (along X direction) of the weld just behind the pin and below the shoulder. Temperature distribution is symmetric for the area occupied

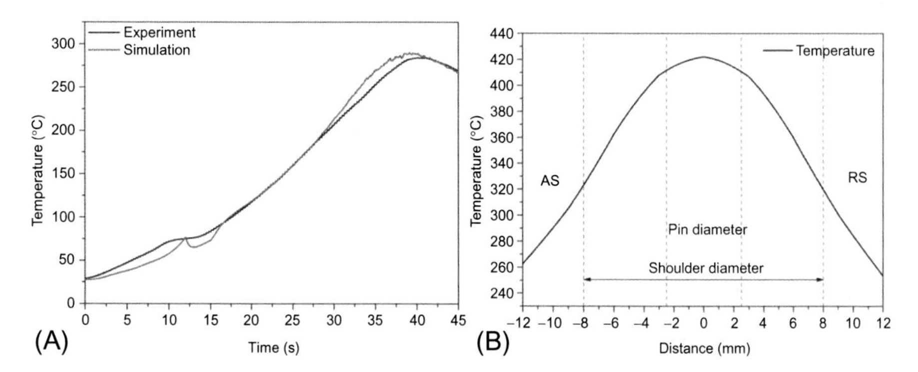

Fig. 5.14 Temperature distribution: (A) validation of model-predicted temperature with experimental and (B) temperature variation along the transverse direction of the weld.

by the pin and afterward a gradual increase of temperature in the AS is observed as compared to the RS. Temperature difference between AS and RS at the edge of pin is less than 1°C; it is 3°C at the edge of shoulder (8 mm from center) and 9°C at a distance of 12 mm from center. This is because of higher relative velocity at the edge of the shoulder due to larger radius as compared with the pin.

5.4.7.2 Evolution of force and torque during FSW

Rigidity of the machine should be such that it can bear the reaction force. On the other hand, spindle torque provides information about the power consumption which will be the deciding factor for the machine capacity. Fig. 5.15A and B show the evolution of axial force and torque during FSW, respectively.

Initial plunge of the tool on the workpiece leads to initial peak in the axial force and afterward it decreases slightly due to rise in temperature of the workpiece due to friction and plastic deformation. Maximum force is observed at the plunging stage when shoulder comes in contact with the workpiece and is impinged further to provide the desired plunge depth. Sudden drop in force is observed after plunging as the material reaches to viscous state. In welding phase, force attains almost a constant value as shown. Average welding force during welding phase is 4900 N. Torque increases continuously during the plunging phase due to increase in power requirement. Maximum torque predicted during this phase is 14.45 N-m. During dwelling, torque requirement drops before attaining a steady value in welding stage. Drop in torque is mainly due to rise in temperature of the workpiece during plunging phase. Average torque during welding phase is 9.5 N-m, which corresponds to power requirement of 895.4 W. This is much lower as compared to some of the fusion welding techniques like laser and electron beam welding.

This methodology is fruitful for developing a complete model for FSW that can predict all the output responses including forces and material flow. Researchers have reported a real-time material flow by using particle tracking method and the same is also used to study the time domain behavior of temperature, strain rate, material velocity, etc. Another characteristic of this method is its ability to predict microstructure by using methods like cellular automata (Asadi et al., 2015) and Monte Carlo simulation (Zhang et al., 2015).

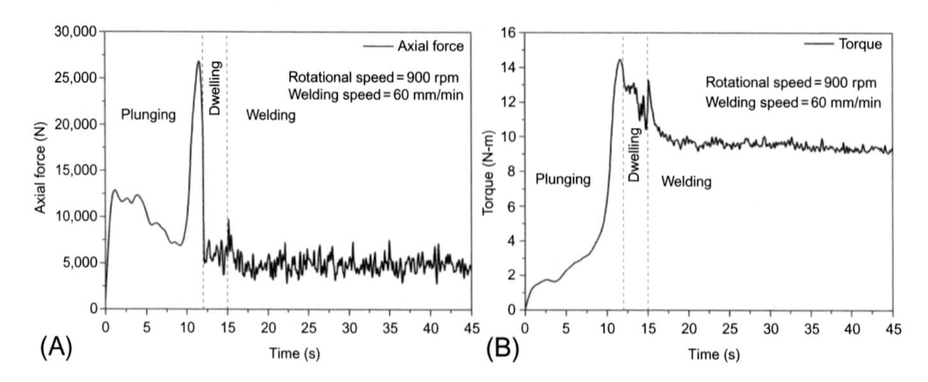

Fig. 5.15 Evolution of (A) axial force and (B) torque during FSW.

5.5 Modeling of FSW with Eulerian analysis

Major problem associated with Lagrangian method is the distortion of the mesh element leading to premature failure of the simulation. This issue can be tackled with the Eulerian analysis. ANSYS Fluent has strong capabilities to model viscous fluid flow. Compared to Lagrangian analysis, this method takes lower time to achieve the convergence. Plunging phase of FSW cannot be simulated with this method and modeling of only dwelling and welding can be performed. Seide and Reynolds (2003) and Colegrove and Shercliff (2004) are some of the initial researchers who worked in the area of modeling of FSW with the Eulerian analysis. Arora et al. developed a code based on Eulerian analysis to calculate the load-bearing capacity of the tool (Arora et al., 2011b), forces, and torque (Arora et al., 2009). Various researchers have used this method to simulate lap welding of dissimilar material (Tang and Shen, 2016) and reverse dual rotation FSW (Li and Liu, 2013). Most of the researchers have selected the coefficient of friction and slip rate based on hit and trial method to validate the model results with experiment. Attempts have been made to develop a methodology to calculate those based on experimental force and torque (Su et al., 2014, 2015). Subsequent section deals with the procedure to model FSW by using Eulerian method in ANSYS Fluent. This methodology is also valid for other software based on Eulerian method but some of the terminology might change.

5.5.1 Geometric modeling and boundary condition

Eulerian analysis considers material as a fluid, and hence defining a solid tool is not feasible. Therefore impression of the tool geometry (shoulder and pin) is created on the workpiece, as shown in Fig. 5.16, either by using Boolean operation in the design modeler of ANSYS Fluent or by creating it in any CAD software and importing the same in the simulation environment. Fig. 5.16B shows the enlarged view of the features of the tool embedded on the workpiece surface. Length and width of workpiece are 110 mm and 114 mm, respectively, with a thickness of 3.1 mm. Tool shoulder diameter is 16 mm with a cylindrical pin of 5 mm diameter and 2.5 mm height. Since it is a steady-state analysis, only welding phase is simulated. All walls (top, bottom, and side) are defined as a stationary wall with a traction-free surface. Welding velocity is defined as inlet flow in positive Y direction and at the outlet atmospheric

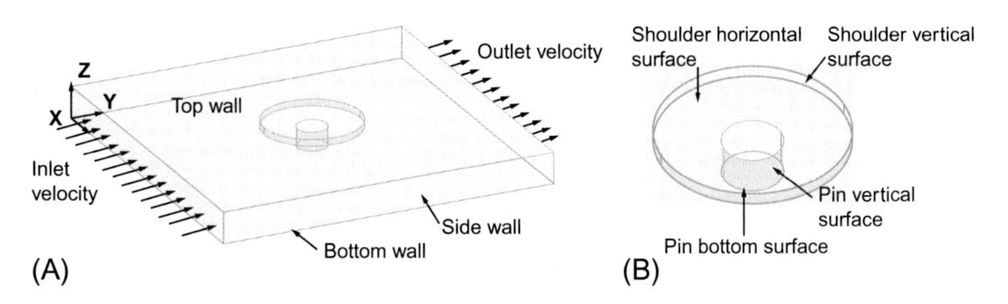

Fig. 5.16 (A) Geometric model and boundary conditions and (B) zoomed view of tool surface.

pressure is defined. Heat partitioning between the workpiece and the tool is defined based on Eq. (5.21).

$$f_w = \frac{\sqrt{(K\rho c_p)_w}}{\sqrt{(K\rho c_p)_w} + \sqrt{(K\rho c_p)_t}} \tag{5.21}$$

where f_w is the percentage of total heat transferred on the workpiece, and subscripts w and t indicate the physical properties of the workpiece and the tool. Therefore 70% of the heat generated is distributed to the workpiece and remaining to the tool. The use of the previous equation is valid when the tool and workpiece are considered as infinite heat sink and there is no influence of heat flow from the tool boundary. Convective heat transfer is defined between workpiece–backing plate and workpiece–environment with a heat transfer coefficient of 1000 and 30 $W/m^2/K$, respectively.

5.5.2 Material model

In this analysis, the workpiece is defined as non-Newtonian, incompressible body. Maximum shear yield stress is calculated based on von Mises yield criterion as shown in Eq. (5.22).

$$\tau_y = \frac{\sigma_y}{\sqrt{3}} \tag{5.22}$$

In FSW, deformation takes place if stress is above yield stress and hence yield stress (σ_y) can be approximated to the flow stress of the material. Flow stress of the material is defined as a function of strain rate and temperature by using Sheppard and Wright model, as expressed in Eq. (5.2) and material constants of the equation are summarized in Table 5.8. Calculations of effective strain rate and strain rate component are given by Eqs. (5.15) and (5.16), respectively. It has been reported that Eq. (5.2) becomes inapplicable near the melting point and beyond. Therefore a softening term is added to the equation as shown in Eq. (5.23) (Su et al., 2014).

$$\sigma_e = \left(1 - \sqrt{\frac{T - 273}{T_{\text{melt}} - 273}}\right) \cdot \overline{\sigma} + \sigma_0 \tag{5.23}$$

where σ_e is effective flow stress and σ_0 is the stress of the material above its melting point and is taken as 20MPa. The physical properties of the material are defined in Table 5.12. The viscosity of the material is determined by using effective flow stress and strain using the Perzyna's viscoplasticity model, as described by Eq. (5.24).

$$v = \frac{\sigma_e}{3\dot{\overline{\varepsilon}}} \tag{5.24}$$

where v is the viscosity of the material.

Table 5.12 **Physical properties of AA6061 (Al-Badour et al., 2013)**

Temperature (°C)	Density (kg/m^3)	Specific heat (J/kg/°C)	Young's modulus (GPa)	Poisson's ratio	Thermal conductivity (W/m/°C) (Chao et al., 2008)
25	2.69	945	66.94	0.33	162
100	2.69	978	63.21	0.334	177
149	2.67	1000	61.32	0.335	184
204	2.66	1030	56.80	0.336	192
260	2.66	1052	51.15	0.338	201
316	2.63	1080	47.17	0.36	207
371	2.63	1100	43.51	0.4	217
427	2.60	1130	28.77	0.41	229

5.5.3 Governing equation

Continuity, momentum, and energy equations are the conservation equations in Eulerian analysis as specified as follows:

Continuity equation

$$\frac{\partial u}{\partial x}+\frac{\partial v}{\partial y}+\frac{\partial w}{\partial z}=0 \tag{5.25}$$

Momentum equation

$$\rho\left(u\frac{\partial u}{\partial x}+v\frac{\partial u}{\partial y}+w\frac{\partial u}{\partial z}\right)=-\frac{\partial p}{\partial x}+\frac{\partial}{\partial x}\left(\mu\frac{\partial u}{\partial x}\right)+\frac{\partial}{\partial y}\left(\mu\frac{\partial u}{\partial y}\right)+\frac{\partial}{\partial z}\left(\mu\frac{\partial u}{\partial z}\right) \tag{5.26}$$

$$\rho\left(u\frac{\partial v}{\partial x}+v\frac{\partial v}{\partial y}+w\frac{\partial v}{\partial z}\right)=-\frac{\partial p}{\partial y}+\frac{\partial}{\partial x}\left(\mu\frac{\partial v}{\partial x}\right)+\frac{\partial}{\partial y}\left(\mu\frac{\partial v}{\partial y}\right)+\frac{\partial}{\partial z}\left(\mu\frac{\partial v}{\partial z}\right) \tag{5.27}$$

$$\rho\left(u\frac{\partial w}{\partial x}+v\frac{\partial w}{\partial y}+w\frac{\partial w}{\partial z}\right)=-\frac{\partial p}{\partial z}+\frac{\partial}{\partial x}\left(\mu\frac{\partial w}{\partial x}\right)+\frac{\partial}{\partial y}\left(\mu\frac{\partial w}{\partial y}\right)+\frac{\partial}{\partial z}\left(\mu\frac{\partial w}{\partial z}\right) \tag{5.28}$$

Energy equation

$$\rho C_p\left(u\frac{\partial T}{\partial x}+v\frac{\partial T}{\partial y}+w\frac{\partial T}{\partial z}\right)=\frac{\partial}{\partial x}\left(k\frac{\partial T}{\partial x}\right)+\frac{\partial}{\partial y}\left(k\frac{\partial T}{\partial y}\right)+\frac{\partial}{\partial z}\left(k\frac{\partial T}{\partial z}\right)+S_t \tag{5.29}$$

where u, v, and w are the velocity in X, Y, and Z directions, respectively, and S_t is the viscous heat dissipation. It has been reported that viscous heat dissipation is small as compared to the plastic heat dissipation (Nandan et al., 2008) and is given by Eq. (5.30).

$$S_t=\alpha\phi \tag{5.30}$$

where α is a constant reflecting the extent of mixing and it is defined as 0.05. ϕ is the viscous dissipation heat as shown in Eq. (5.31).

$$\phi=\mu\left[2\left(\frac{\partial u}{\partial x}\right)^2+2\left(\frac{\partial v}{\partial y}\right)^2+2\left(\frac{\partial w}{\partial z}\right)^2+\left(\frac{\partial v}{\partial x}+\frac{\partial u}{\partial y}\right)^2+\left(\frac{\partial v}{\partial z}+\frac{\partial w}{\partial y}\right)^2+\left(\frac{\partial w}{\partial x}+\frac{\partial u}{\partial z}\right)^2\right] \tag{5.31}$$

The coupled equations were solved by using pressure-based solver with second-order discretization of momentum and energy equation

5.5.4 Heat generation

The interfacial heat generation due to friction and plastic deformation between the workpiece and the tool is defined (Schmidt et al., 2004; Schmidt and Hattel, 2008) through the following equation.

For shoulder and pin bottom surface

$$q_1 = \left[\delta\eta\tau_y + (1-\delta)\mu_f p_0\right](\omega r - U\sin\theta) \tag{5.32}$$

For pin side surfaces

$$q_2 = \delta\eta\tau_y(\omega r - U\sin\theta) \tag{5.33}$$

5.5.5 Mesh generation and solver

Initially tetrahedral mesh was attempted for meshing the workpiece but it led to non-convergence of the solution due to low mesh quality and high mesh density. Therefore to improve the convergence domain is meshed with a coupled hexahedral element. The mesh has minimum orthogonal quality and skewness of 0.66 and 0.77, respectively. Pressure–velocity coupling is used to calculate the velocity and temperature fields. PRESTO algorithm is used for the interpolation of cell face pressure. Second-order upwind discretization of energy and momentum is defined. Relaxation factor for momentum and pressure is taken as 0.5 and for energy it is 0.95 to dampen the solution and reduce the steep oscillation which leads to divergence of the solution. Convergence criterion was defined at a residual value of 10^{-6}.

5.5.6 Results and discussion

Temperature distribution for rotational speed of 900 rpm and 60 mm/min welding speed is shown in Fig. 5.17A. Maximum temperature observed is 703 K. The temperature obtained in this method is slightly on a higher side as compared to Lagrangian

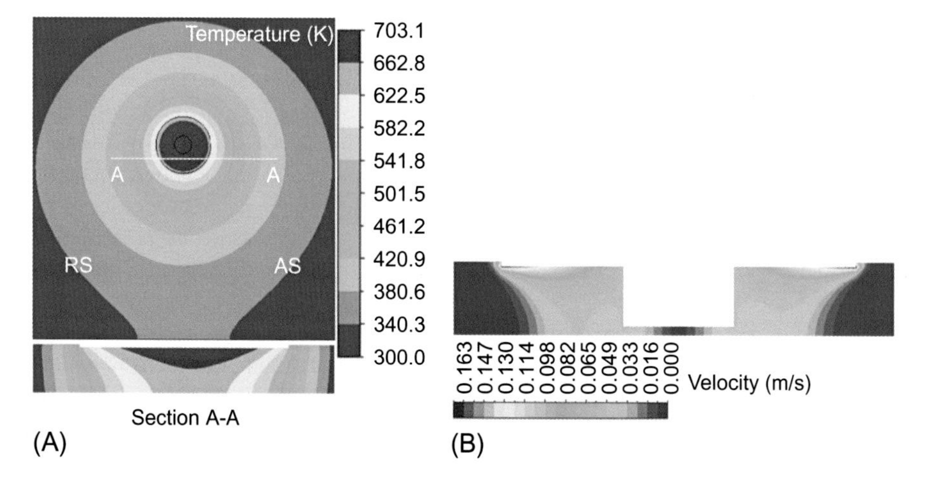

Fig. 5.17 (A) Temperature distribution from top view and transverse section and (B) material velocity distribution in the transverse direction.

analysis and this may be because of difference in frictional contact condition. Temperature contour obtained is similar to Lagrangian analysis. Material velocity distribution is shown in Fig. 5.17B. Maximum velocity is 0.163 m/s at the edge of the shoulder as it has larger radius as compared to the pin. Lower computation cost is the major advantage of Eulerian analysis and it is one of the major reasons why more researchers have opted this method instead of Lagrangian. Also, since temperature in FSW is close to the melting point and material behaves like a viscous fluid, this method is thus considered close to reality. Major drawback of this method is the inability to simulate plunging phase, and quite a few variables such as slip rate, coefficient of friction, etc., need to be iterated to optimize the results.

5.6 Modeling of FSW with coupled Eulerian-Lagrangian (CEL) analysis

This method is a combination of Lagrangian and Eulerian analyses by incorporating the benefits of both. It is based on the principle of VOF that enables it to predict the volumetric defect originating in the process. Only a few researchers have reported the application of CEL method in FSW. Grujicic et al. (2012) have simulated material flow with CEL by defining a marker material within the processed zone. Badour et al. have simulated volumetric defects with CEL method (Al-Badour et al., 2013) and they modeled FSW of AA 6061-AA5083 also (Al-Badour et al., 2014). In the present work a three-dimensional coupled temperature-displacement, dynamic explicit analysis is carried out by using ABAQUS/Explicit based on CEL method.

5.6.1 Geometric modeling and material model

In ABAQUS all parts are created in the part module. Workpiece is defined as a three-dimensional Eulerian body with a dimension of 110 mm × 114 mm × 4.1 mm. The thickness of the workpiece (4.1 mm) is partitioned into two parts to define void surface and material assignment region each having thickness of 1 mm and 3.1 mm, respectively. Void region is required to simulate the flash formation. Tool is modeled as a solid homogenous rigid body with a shoulder diameter of 16 mm and a cylindrical pin with a diameter of 5 mm, and a pin height of 2.5 mm pin. Tool is tilted by 2 degrees toward the trailing edge during the assembly of the parts. AA 6061 is defined as the workpiece material and tool steel H13 is defined as the tool. Physical properties of the material are defined as a function of temperature and are same as presented in Table 5.12. Physical properties of H13 tool steel are mentioned in Table 5.9. Johnson–Cook material model is defined as expressed in Eq. (5.1). Material constant for JC model is presented in Table 5.13.

In CEL analysis if no material property is assigned to a region, it is then considered as void or empty region. Therefore to incorporate the material model and physical properties of the material, material assignment tool is used as shown in Fig. 5.18. The upper part (red color) having a thickness of 1 mm acts like a void region and the lower part (blue color) is the region with assigned material properties. It is

Table 5.13 Material constant for JC model

A (MPa)	B (MPa)	C	m	n	T_{room} (°C)	T_{melt} (°C)
324	114	0.002	1.34	0.42	24	583

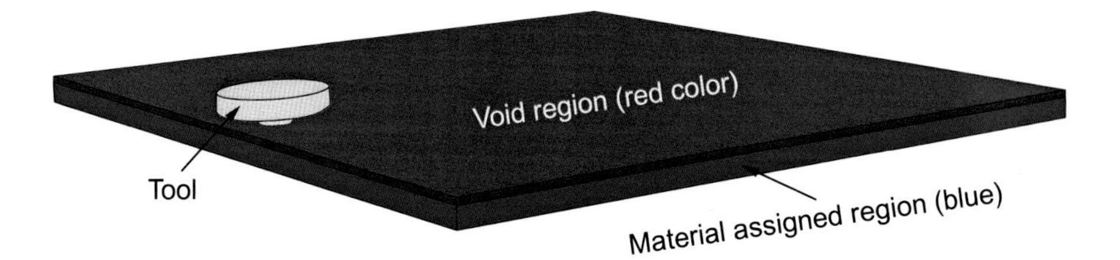

Fig. 5.18 Tool and workpiece assembly indicating void and material assigned region.

mandatory to define a void region or else simulation will not converge because during plunging stage the tool pin will impinge on the workpiece and occupy the volume inside the Eulerian body (material assigned region). Hence, material beneath the pin will flow in upward direction and occupy space in the void region (ABAQUS 6.14 documentation, 2015).

5.6.2 Mesh generation

An eight-noded thermally coupled linear Eulerian brick, reduced integration, hourglass control (EC3D8RT) is defined to mesh the workpiece. Finer mesh (element size of 0.4 mm) is defined near the interaction zone of tool and the workpiece by partitioning the workpiece, as shown in Fig. 5.19.

Rest of the region is discretized with a mesh size of 4 mm. Workpiece is meshed with 77,325 number of elements. Tool is meshed with an eight-noded thermally coupled brick trilinear displacement and temperature reduced integration (C3D8RT). Tool is meshed with a mesh size of 0.5 mm leading to 7672 number of elements.

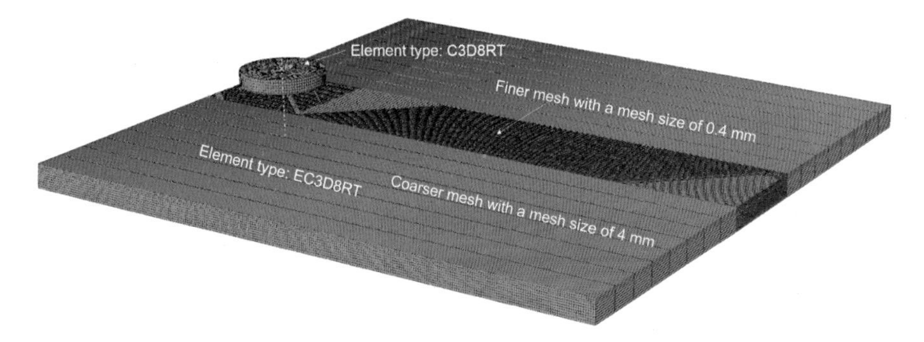

Fig. 5.19 Meshed assembly of the tool and workpiece.

5.6.2.1 Mesh refinement study

Mesh size influences numerical accuracy as well as the computation time. Mesh refinement study has been conducted to study the effect of mesh on accuracy and computation time. The effect of mesh size on axial force and temperature was analyzed as shown in Fig. 5.20A and B. Also the effect of mesh size on computation time is presented in Table 5.14.

Fluctuation of axial force is observed for 0.6 mm during plunging phase. Mesh size of 0.6 mm and 0.4 predicted higher maximum temperature compared to the other two mesh sizes with a deviation of 6%, but at the same time mesh size of 0.4 mm takes 35% lower computational time as compared with 0.3 mm mesh size. Therefore based on the computation cost and lower deviation in results, mesh size of 0.4 mm is chosen for the analysis.

5.6.3 Contact interaction

General contact is defined between the tool and workpiece to incorporate all contacting pairs including the self-contact of the workpiece. Coulomb's law of friction is used to define the tangential traction with a coefficient of friction of 0.8. Penalty contact algorithm is defined for the friction formulation. Hard contact pressure enclosure is used to define the normal behavior at the interface. This method minimizes the penetration of the slave surface into master surface at the constraint location and does not allow the transfer of tensile stress across the interface. Contact is separated wherever tensile stress is produced and remains intact for compressive stress. Heat generation due to friction and plastic deformation is considered as established in Eqs. (5.6)–(5.9). It is assumed that 100% of frictional energy is converted to heat, and inelastic heat fraction is defined as 0.9, indicating 90% of the plastic deformation is converted to heat. Heat generated at the interface is transferred to the tool, workpiece, backing plate, and environment. Heat partitioning between the tool and workpiece is calculated based on Eq. (5.21) and 70% of total heat is transferred to the workpiece and 30% on the tool. Convective heat transfer is defined to consider heat transfer between workpiece–backing plate and workpiece–environment, as expressed

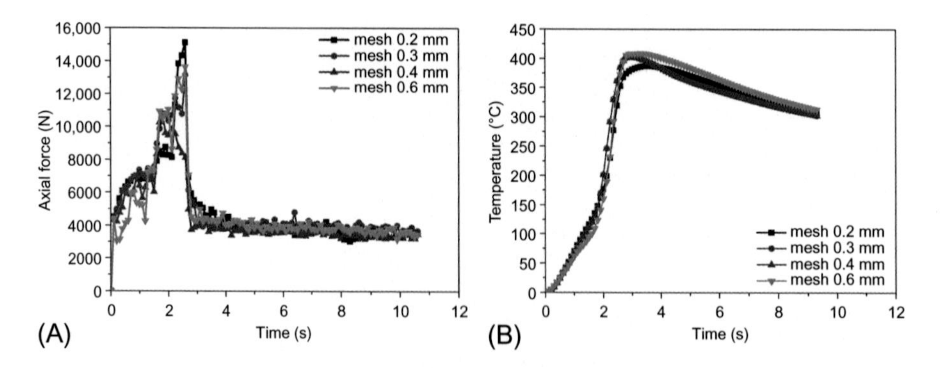

Fig. 5.20 Influence of mesh size on (A) axial force and (B) temperature.

Table 5.14 Influence of mesh size on computational time

Mesh size (mm)	No. of elements	Computational time (h)
0.2	455,566	61
0.3	150,000	27
0.4	77,325	20
0.6	56,516	4

in Eqs. (5.10) and (5.11), respectively. Convective heat transfer coefficients for the former and the latter are defined as 4000 and 25 W/m^2°C, respectively.

5.6.4 Boundary conditions

Various boundary conditions are required to replicate the FSW in simulation. Boundary conditions need to be defined based on the FSW stage, as shown in Fig. 5.21.

During plunging stage, tool is given rotational and travel velocities along negative Z direction. Velocities at the bottom, side, and front face of workpiece are arrested in *Z*, *Y*, and *X* directions, respectively, as shown in Fig. 5.21 to avoid spilling of Eulerian material from the domain. During welding phase, all boundary conditions discussed for plunging are valid with two changes. First, linear velocity defined on the tool and zero velocity (V_x) defined on front face of the workpiece are removed, and second welding speed is defined in *X* direction based on inflow and outflow of the material as shown in Fig. 5.21.

5.6.5 Governing equations

Conservation equations for the Eulerian analysis are spatial time derivative while for Lagrangian it is material time derivative. The relation between spatial and material time derivative is expressed in Eqs. (5.34) (Benson and Okazawa, 2004)

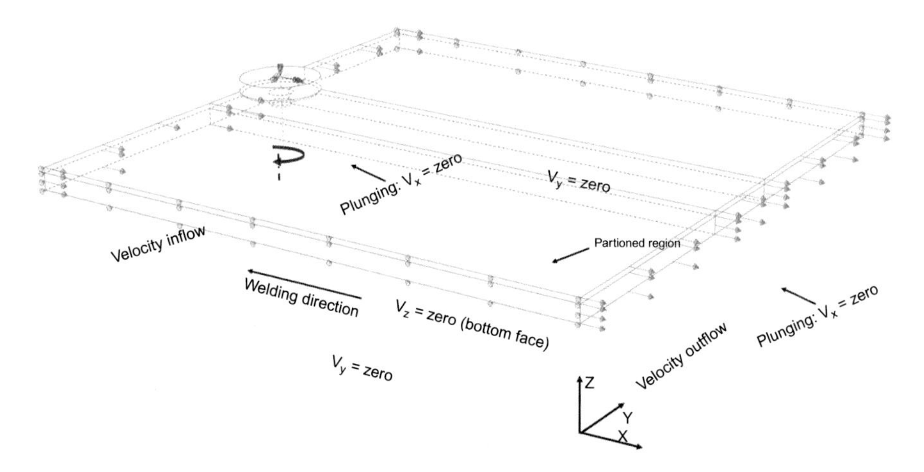

Fig. 5.21 Boundary condition for different stage of FSW.

$$\frac{D\phi}{Dt}=\frac{\partial\phi}{\partial t}+u\cdot(\nabla\phi) \tag{5.34}$$

where ϕ is the arbitrary solution variable and u is the material velocity. $\frac{D\phi}{Dt}$ and $\frac{\partial\phi}{\partial t}$ are material and spatial time derivatives of ϕ, respectively. The Lagrangian mass, momentum, and energy equations are transformed into Eulerian conservation (spatial derivative) (Benson and Okazawa, 2004) and are given by Eqs. (5.35), (5.36), and (5.37), respectively.

$$\frac{\partial\rho}{\partial t}+u\cdot(\nabla\rho)+\rho\nabla\cdot u=0 \tag{5.35}$$

$$\frac{\partial u}{\partial t}+u\cdot(\nabla\cdot u)=\frac{1}{\rho}(\nabla\cdot\sigma)+b \tag{5.36}$$

$$\frac{\partial e}{\partial t}+u\cdot(\nabla e)=\sigma:D \tag{5.37}$$

where b and e are body force and internal energy. Furthermore, the previous three Eulerian equations can be arranged into the conservative forms, as shown in Eqs. (5.38)–(5.40).

$$\frac{\partial\rho}{\partial t}+\nabla\cdot(\rho u)=0 \tag{5.38}$$

$$\frac{\partial(\rho u)}{\partial t}+\nabla\cdot(\rho u\otimes u)=\nabla\cdot\sigma+\rho b \tag{5.39}$$

$$\frac{\partial e}{\partial t}+\nabla\cdot(eu)=\sigma:D \tag{5.40}$$

The Eulerian governing equation discussed in Eqs. (5.38)–(5.40) has a general conservation form, as shown in Eq. (5.41).

$$\frac{\partial\phi}{\partial t}+\nabla\cdot\Phi=S \tag{5.41}$$

where Φ is flux function and S is the source term. Operator splitting divides Eq. (5.41) into two equations as shown in Eqs. (5.42) and (5.43), and these two equations are solved sequentially in CEL method

$$\frac{\partial\phi}{\partial t}=S \tag{5.42}$$

$$\frac{\partial\phi}{\partial t}+\nabla\cdot\Phi=0 \tag{5.43}$$

Eq. (5.42) is the Lagrangian step containing the source term and Eq. (5.43) is the Eulerian step containing the convective term. Schematic of split operator for each step of CEL method is shown in Fig. 5.22.

Eq. (5.42) becomes identical to the standard Lagrangian method if the spatial time derivative is changed to the material time derivative on the fixed mesh. Eq. (5.43) is solved by moving the deformed mesh back to the original one and volume of the material is transported to original mesh using any transportation method. The Lagrangian solution variables, viz., mass, energy, stress, etc., are adjusted to account for the flow of the material by transport algorithm (Benson and Okazawa, 2004). Further detail on the derivation of governing equation and methodology can be found in Benson (1992, 1997).

5.6.6 Results and discussion

(Fig. 5.23A shows the temperature distribution at the top surface and in the transverse direction. The maximum temperature observed is 568°C which is 0.95 times the melting point of the material in degree centigrade. The predicted temperature is on a higher side as compared to the Lagrangian analysis due to higher chosen coefficient of friction. Fig. 5.23B compares the model-predicted and experimental axial force for during

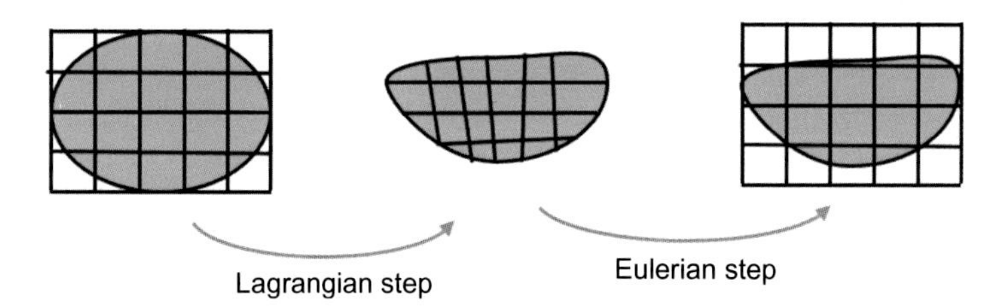

Fig. 5.22 Schematic of split operator for CEL method.

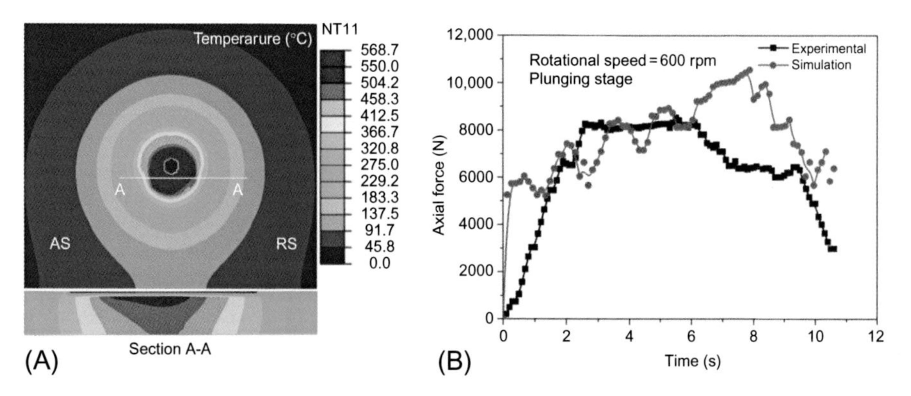

Fig. 5.23 (A) Temperature distribution at 900 rpm and 60 mm/min and (B) validation of model-predicted force with experiment at 600 rpm.

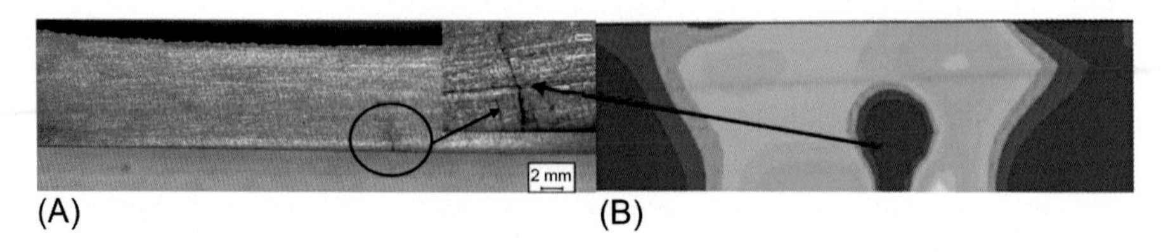

Fig. 5.24 Defect prediction: (A) experimental and (B) CEL method for 1500 rpm and 60 mm/min.

plunging stage for a rotational speed of 600 rpm and plunge rate of 15 mm/min. The trend of both the curves is similar but simulation result has predicted slightly higher axial force as compared to experiment.

Root flaw defect is observed for a rotational speed of 1500 rpm and 60 mm/min as shown in the macrograph of the aluminum sample in Fig. 5.24A. Fig. 5.24B shows the plastic strain distribution for the same parameters. At the bottom of the weld, the model has predicted zero plastic strain due to absence of the material in that region, hence successfully predicting the root flaw defect.

5.7 Comparison of modeling methods

Three different methodologies discussed for the modeling of FSW are distinct in their own ways. Each of them is capable of modeling FSW but to validate the results friction condition and heat transfer coefficient need to be optimized. Temperature results predicted by Lagrangian and Eulerian method are close to reality while CEL method predicts higher maximum temperature that is very much close to the melting point of the material. This is due the JC material model used in CEL method which is unable to predict the saturation temperature. Among the three, only CEL method can predict the volumetric defect due to application of VOF principle. CEL method requires superior computation workstation while other two can be performed on normal computers. Lagrangian method implemented through DEFORM-3D can predict microstructure and grain size of the weld which gives it an edge over the other two methods.

5.8 Conclusion

Three different methodologies have been discussed for the modeling of friction stir welding process based on Lagrangian, Eulerian, and coupled Lagrangian and Eulerian principle. Each of the discussed method is capable of modeling FSW and has its own advantages and disadvantages. Lagrangian method is capable of predicting all the output responses and can be extended to predict grain size and microstructure of the material. But this method has to be used along with a strong remeshing technique. Eulerian analysis is the least time consuming among the three but cannot simulate the plunging stage. Also, only streamline material flow is predicted through this method. CEL method captures the advantages of both the methods and is capable of handling large deformation process. Ability to predict volumetric defects makes this method unique. Though this method is the most computationally expensive but that can be used with mass scaling method.

References

ABAQUS 6.14 documentation, 2015.

Al-Badour, F., Merah, N., Shuaib, A., Bazoune, A., 2013. Coupled Eulerian Lagrangian finite element modeling of friction stir welding processes. J. Mater. Process. Technol. 213, 1433–1439. http://dx.doi.org/10.1016/j.jmatprotec.2013.02.014.

Al-Badour, F., Merah, N., Shuaib, A., Bazoune, A., 2014. Thermo-mechanical finite element model of friction stir welding of dissimilar alloys. Int. J. Adv. Manuf. Technol. 72, 607–617. http://dx.doi.org/10.1007/s00170-014-5680-3.

Alfaro, I., Racineux, G., Poitou, A., Cueto, E., Chinesta, F., 2009. Numerical simulation of friction stir welding by natural element methods. Int. J. Mater. Form. 2, 225–234. http://dx.doi.org/10.1007/s12289-009-0406-z.

Arora, A., Nandan, R., Reynolds, A.P., DebRoy, T., 2009. Torque, power requirement and stir zone geometry in friction stir welding through modeling and experiments. Scr. Mater. 60, 13–16. http://dx.doi.org/10.1016/j.scriptamat.2008.08.015.

Arora, A., De, A., Debroy, T., 2011a. Toward optimum friction stir welding tool shoulder diameter. Scr. Mater. 64, 9–12. http://dx.doi.org/10.1016/j.scriptamat.2010.08.052.

Arora, A., Mehta, M., De, A., DebRoy, T., 2011b. Load bearing capacity of tool pin during friction stir welding. Int. J. Adv. Manuf. Technol. 61, 911–920. http://dx.doi.org/10.1007/s00170-011-3759-7.

Asadi, P., Mahdavinejad, R.A., Tutunchilar, S., 2011. Simulation and experimental investigation of FSP of AZ91 magnesium alloy. Mater. Sci. Eng. A. 528, 6469–6477. http://dx.doi.org/10.1016/j.msea.2011.05.035.

Asadi, P., Besharati Givi, M.K., Akbari, M., 2015. Simulation of dynamic recrystallization process during friction stir welding of AZ91 magnesium alloy. Int. J. Adv. Manuf. Technol. http://dx.doi.org/10.1007/s00170-015-7595-z.

Assidi, M., Fourment, L., Guerdoux, S., Nelson, T., 2010. Friction model for friction stir welding process simulation: calibrations from welding experiments. Int. J. Mach. Tools Manuf. 50, 143–155. http://dx.doi.org/10.1016/j.ijmachtools.2009.11.008.

Benson, D.J., 1992. Computational methods in Lagrangian and Eulerian hydrocodes. Comput. Methods Appl. Mech. Eng. 99, 235–394. http://dx.doi.org/10.1016/0045-7825(92)90042-I.

Benson, D.J., 1997. A mixture theory for contact in multi-material Eulerian formulations. Comput. Methods Appl. Mech. Eng. 140, 59–86. http://dx.doi.org/10.1016/S0045-7825(96)01050-X.

Benson, D.J., Okazawa, S., 2004. Contact in a multi-material Eulerian finite element formulation. Comput. Methods Appl. Mech. Eng. 193, 4277–4298. http://dx.doi.org/10.1016/j.cma.2003.12.061.

Biswas, P., Mandal, N.R., 2011. Effect of tool geometries on thermal history of FSW of AA1100. Weld. J., 129–135.

Buffa, G., Fratini, L., 2009. Friction stir welding of steels: process design through continuum based FEM model. Sci. Technol. Weld. Join. 14, 239–246. http://dx.doi.org/10.1179/136217109X421328.

Buffa, G., Hua, J., Shivpuri, R., Fratini, L., 2006a. A continuum based fem model for friction stir welding—model development. Mater. Sci. Eng. A. 419, 389–396. http://dx.doi.org/10.1016/j.msea.2005.09.040.

Buffa, G., Hua, J., Shivpuri, R., Fratini, L., 2006b. Design of the friction stir welding tool using the continuum based FEM model. Mater. Sci. Eng. A. 419, 381–388. http://dx.doi.org/10.1016/j.msea.2005.09.041.

Buffa, G., Baffari, D., Di Caro, A., Fratini, L., 2015. Friction stir welding of dissimilar aluminium–magnesium joints: sheet mutual position effects. Sci. Technol. Weld. Join. 20 (, 271–279. http://dx.doi.org/10.1179/1362171815Y.0000000016.

Chao, Y.J., Liu, S., Chien, C., 2008. Friction stir welding of al 6061-T6 thick plates: Part II - numerical modeling of the thermal and heat transfer phenomena. J. Chinese Inst. Eng. 31, 769–779. http://dx.doi.org/10.1080/02533839.2008.9671431.

Colegrove, P.A., Shercliff, H.R., 2004. Development of Trivex friction stir welding tool Part 1 – two-dimensional flow modelling and experimental validation. Sci. Technol. Weld. Join. 9, 352–361. http://dx.doi.org/10.1179/136217104225021661.

Das, R., Cleary, P.W., 2016. Three-dimensional modelling of coupled flow dynamics, heat transfer and residual stress generation in arc welding processes using the mesh-free SPH method. J. Comput. Sci. 16, 200–216. http://dx.doi.org/10.1016/j.jocs.2016.03.006.

DEFORM v 11.0 documentation, 2015.

Donea, J., Huerta, A., Ponthot, J., Rodr, A., 1999. Arbitrary Lagrangian – Eulerian Methods. In: Stein, E., de Borst, R., Hughes, T.J.R. (Eds.), In: Encycl. Comput. Mech., John Wiley and sons, pp. 1–25. http://dx.doi.org/10.1002/0470091355.ecm009.

Fratini, L., Buffa, G., Lo Monaco, L., 2010. Improved FE model for simulation of friction stir welding of different materials. Sci. Technol. Weld. Join. 15, 199–207. http://dx.doi.org/10.1179/136217110X12665048207575.

Grujicic, M., Arakere, G., Pandurangan, B., Ochterbeck, J.M., Yen, C.F., Cheeseman, B.A., et al., 2012. Computational analysis of material flow during friction stir welding of AA5059 aluminum alloys. J. Mater. Eng. Perform. 21, 1824–1840. http://dx.doi.org/10.1007/s11665-011-0069-z.

Grujicic, M., Ramaswami, S., Snipes, J.S., Avuthu, V., Galgalikar, R., Zhang, Z., 2015. Prediction of the grain-microstructure evolution within a friction stir welding (FSW) joint via the use of the Monte Carlo simulation method. J. Mater. Eng. Perform. 24, 3471–3486. http://dx.doi.org/10.1007/s11665-015-1635-6.

Jain, R., Pal, S.K., Singh, S.B., 2014. Finite element simulation of temperature and strain distribution in Al2024 aluminum alloy by friction stir welding. In: 5th Int. 26th All India Manuf. Technol. Des. Res. Conf. (AIMTDR 2014), 2014, pp. 3–7.

Jain, R., Kumari, K., Kesharwani, R.K., Kumar, S., Pal, S.K., Singh, S.B., et al., 2015. Friction stir welding: scope and recent development. In: Paulo Davim, J. (Ed.), In: Mordern Manuf. Eng. Ed, pp. 179–228. http://dx.doi.org/10.1007/978-3-319-20152-8.

Jain, R., Pal, S.K., Singh, S.B., 2016a. A study on the variation of forces and temperature in a friction stir welding process: a finite element approach. J. Manuf. Process. 23, 278–286. http://dx.doi.org/10.1016/j.jmapro.2016.04.008.

Jain, R., Pal, S.K., Singh, S.B., 2016b. Finite element simulation of temperature and strain distribution in Al2024 aluminum alloy by friction stir welding. J. Inst. Eng. Ser. C. http://dx.doi.org/10.13140/RG.2.1.2429.8408.

Jain, R., Pal, S.K., Singh, S.B., 2017. Finite element simulation of pin shape influence on material flow, forces in friction stir welding. Int. J. Adv. Manuf. Technol. http://dx.doi.org/10.1007/s00170-017-0215-3.

Johnson, G.R., Cook, W.H., 1983. A constitutive model and data for metals subjected to large strains, high strain rates and high temperatures. Proc. 7th Int. Symp. Ballist. 547, 541–547. http://dx.doi.org/10.1038/nrm3209.

Kheireddine, A.H., Khalil, A.A., Ammouri, A.H., Kridli, G.T., Hamade, R.F., Engineering, M.S., 2013. An experimentally validated thermo mechanical finite element model for friction stir welding in carbon steels. In: ICAIM, pp. 2–5.

Kobayashi, S., Oh, S.-I., Altan, T., 1989. Metal forming and the finite element method. Oxford series, New York.

Kuykendall, K., 2011. An Evaluation of Constitutive Laws and their Ability to Predict Flow Stress over Large Variations in Temperature, Strain, and Strain Rate Characteristic of Friction Stir Welding. Brigham Young University.

Kuykendall, K., Nelson, T., Sorensen, C., 2013. On the selection of constitutive laws used in modeling friction stir welding. Int. J. Mach. Tools Manuf. 74, 74–85. http://dx.doi.org/10.1016/j.ijmachtools.2013.07.004.

Laboratory, N.P., 2007. Manual for the Calculation of Elastic-Plastic Materials Models Parameters. National Physical laboratory Queens Printer, Scotland, UK

Li, J.Q., Liu, H.J., 2013. Characteristics of the reverse dual-rotation friction stir welding conducted on 2219-T6 aluminum alloy. Mater. Des. 45, 148–154. http://dx.doi.org/10.1016/j.matdes.2012.08.068.

Marzbanrad, J., Akbari, M., Asadi, P., Safaee, S., 2014. Characterization of the Influence of tool pin profile on microstructural and mechanical properties of friction stir welding. Metall. Mater. Trans. B Process Metall. Mater. Process. Sci. 45, 1887–1894. http://dx.doi.org/10.1007/s11663-014-0089-9.

Mishra, R.S., Ma, Z.Y., 2005. Friction stir welding and processing. Mater. Sci. Eng. R Reports 50, 1–78. http://dx.doi.org/10.1016/j.mser.2005.07.001.

Mishra, R.S., De, P.S., Kumar, N., 2014. Friction Stir Welding and Processing Science and Engineering. Springer International Publishing. http://dx.doi.org/10.1007/978-3-319-07043-8.

Nandan, R., Roy, G.G., Debroy, T., Introduction, I., 2006. Numerical simulation of three-dimensional heat transfer and plastic flow during friction stir welding. Metall. Mater. Trans. A. 37, 1247–1259. http://dx.doi.org/10.1007/s11661-006-1076-9.

Nandan, R., Roy, G.G., Lienert, T.J., Debroy, T., 2007. Three-dimensional heat and material flow during friction stir welding of mild steel. Acta Mater. 55, 883–895. http://dx.doi.org/10.1016/j.actamat.2006.09.009.

Nandan, R., Debroy, T., Bhadeshia, H., 2008. Recent advances in friction-stir welding—process, weldment structure and properties. Prog. Mater. Sci. 53, 980–1023. http://dx.doi.org/10.1016/j.pmatsci.2008.05.001.

Pashazadeh, H., Teimournezhad, J., Masoumi, A., 2014. Numerical investigation on the mechanical, thermal, metallurgical and material flow characteristics in friction stir welding of copper sheets with experimental verification. Mater. Des. 55, 619–632. http://dx.doi.org/10.1016/j.matdes.2013.09.028.

Penalty Contact Algorithm.pdf, n.d. http:/web.mit.edu/calculix_v2.7/CalculiX/ccx_2.7/doc/ccx/node111.html.

Priyadarshini, A., Pal, S.K., Samantaray, A.K., 2012. Finite element modeling of chip formation in orthogonal machining. Stat. Comput. Tech. Manuf., 101–144. http://dx.doi.org/10.1007/978-3-642-25859-6_3.

Rout, M., Pal, S.K., Singh, S.B., 2016. Finite element simulation of a cross rolling process. J. Manuf. Process. 24, 283–292.

Schmidt, H., Hattel, J., 2004. A local model for the thermomechanical conditions in friction stir welding. Model. Simul. Mater. Sci. Eng. 13, 77–93. http://dx.doi.org/10.1088/0965-0393/13/1/006.

Schmidt, H.B., Hattel, J.H., 2008. Thermal modelling of friction stir welding. Scr. Mater. 58, 332–337. http://dx.doi.org/10.1016/j.scriptamat.2007.10.008.

Schmidt, H., Hattel, J., Wert, J., 2004. An analytical model for the heat generation in friction stir welding. Model. Simul. Mater. Sci. Eng. 12, 143–157. http://dx.doi.org/10.1088/0965-0393/12/1/013.

Seide, T.U., Reynolds, A.P., 2003. Two-dimensional friction stir welding process model based on fluid mechanics. Sci. Technol. Weld. Join. 8, 175–183. http://dx.doi.org/10.1179/136217103225010952.

Sellars, C.M., Tegart, W.J.M., 1972. Hot workability. Int. Metall. Rev. 17, 1–24. http://dx.doi.org/10.1179/095066072790137765.

Sheppard, T., Jackson, A., 1997. Constitutive equations for use in prediction of flow stress during extrusion of aluminium alloys. Mater. Sci. Technol. 13, 203–209. http://dx.doi.org/10.1179/026708397790302476.

Sheppard, T., Wright, D.S., 1979. Determination of flow stress: part 1 constitutive equation for aluminum alloys at elevated temperatures. Met. Technol. (June), 215–223.

Soundararajan, V., Zekovic, S., Kovacevic, R., 2005. Thermo-mechanical model with adaptive boundary conditions for friction stir welding of Al 6061. Int. J. Mach. Tools Manuf. 45, 1577–1587. http://dx.doi.org/10.1016/j.ijmachtools.2005.02.008.

Su, H., Wu, C.S., Pittner, A., Rethmeier, M., 2014. Thermal energy generation and distribution in friction stir welding of aluminum alloys. Energy 77, 720–731. http://dx.doi.org/10.1016/j.energy.2014.09.045.

Su, H., Wu, C.S., Bachmann, M., Rethmeier, M., 2015. Numerical modeling for the effect of pin profiles on thermal and material flow characteristics in friction stir welding. Mater. Des. 77, 114–125. http://dx.doi.org/10.1016/j.matdes.2015.04.012.

Tang, J., Shen, Y., 2016. Numerical simulation and experimental investigation of friction stir lap welding between aluminum alloys AA2024 and AA7075. J. Alloys Compd. 666, 493–500. http://dx.doi.org/10.1016/j.jallcom.2016.01.138.

Thomas, W.M., Nicholas, E.D., Needham, J.C., Murch, M.G., Temple Smith, P., Dawes, C.J., 1991. International patent number PCT/GB92/02203 and GB patent application number 9125978.9.

Trimble, D., Monaghan, J., O'Donnell, G.E., 2012. Force generation during friction stir welding of AA2024-T3. CIRP Ann. Manuf. Technol. 61, 9–12. http://dx.doi.org/10.1016/j.cirp.2012.03.024.

Tutunchilar, S., Haghpanahi, M., Besharati Givi, M.K., Asadi, P., Bahemmat, P., 2012. Simulation of material flow in friction stir processing of a cast Al-Si alloy. Mater. Des. 40, 415–426. http://dx.doi.org/10.1016/j.matdes.2012.04.001.

Uyyuru, R.K., Kailas, S.V., 2006. Numerical analysis of friction stir welding process. J. Mater. Eng. Perform. 15, 505–518. http://dx.doi.org/10.1361/105994906X136070.

Xu, S., Deng, X., Reynolds, A.P., Seidel, T.U., 2001. Finite element simulation of material flow in friction stir welding. Sci. Technol. Weld. Join. 6, 191–193. http://dx.doi.org/10.1179/136217101101538640.

Yu, M., Li, W.Y., Li, J.L., Chao, Y.J., 2012. Modelling of entire friction stir welding process by explicit finite element method. Mater. Sci. Technol. 28, 812–817. http://dx.doi.org/10.1179/1743284711Y.0000000087.

Zhang, Z., Zhang, H.W., 2009. Numerical studies on the effect of transverse speed in friction stir welding. Mater. Des. 30, 900–907. http://dx.doi.org/10.1016/j.matdes.2008.05.029.

Zhang, H.W., Zhang, Z., Chen, J.T., 2007. 3D modeling of material flow in friction stir welding under different process parameters. J. Mater. Process. Technol. 183, 62–70. http://dx.doi.org/10.1016/j.jmatprotec.2006.09.027.

Zhang, Z., Wu, Q., Grujicic, M., Wan, Z.Y., 2015. Monte Carlo simulation of grain growth and welding zones in friction stir welding of AA6082-T6. J. Mater. Sci. http://dx.doi.org/10.1007/s10853-015-9495-x.

Zhao, Y., Liu, H., Yang, T., Lin, Z., Hu, Y., 2015. Study of temperature and material flow during friction spot welding of 7B04-T74 aluminum alloy. Int. J. Adv. Manuf. Technol. http://dx.doi.org/10.1007/s00170-015-7681-2.

Modeling of hard machining

6

N.E. Karkalos, A.P. Markopoulos
National Technical University of Athens, Athens, Greece

6.1 Introduction to hard machining

Hard machining can be considered as a special category of general machining processes, aiming at the machining of hardened steel workpieces. This subcategory of machining exhibits several particularities that can seriously affect the process outcome at every aspect, e.g., workpiece surface quality, cutting tool wear, and subsequently cost of manufacture of hardened parts. For this reason, research is being conducted on the determination of optimal machining conditions, minimization of process costs, increase of process efficiency, as well as to explain the mechanism of hard machining processes. For that reason, simulations using well-established numerical and statistical methods are required to support the findings of experimental work and provide useful insights into the complex phenomena occurring during hard machining. In this section of the chapter, an introduction to hard machining is attempted and several notable categories of hard machining are presented, see Fig. 6.1, as well as some of the phenomena occurring during hard machining.

6.1.1 Hard turning

Hard turning is considered as a relatively recent technology for the machining of hardened steels, developed initially in the 1970s. This process is generally accepted to be employed for workpiece materials with hardness over 45 HRC (Dogra et al., 2010) and typically in the range of 45–65 HRC (Campos et al., 2014). Grzesik (2008) considers hard turning process to be the dominant machining operation for hardened materials. The classification of hard materials includes mainly hardened steels, high-speed steels, tool steels, and bearing steels as well as Inconel and alloys suitable for biomedical and other special applications. It is worth noting that some researchers reject the inclusion of the latter mentioned materials in the general category of hard machining materials as their hardness in fact may lie below the range 45–65 HRC (Astakhov, 2011; Campos et al., 2014). Notable applications of hard turning final products in the industry include ball bearings, cam shafts, gears and other related components, automotive applications such as transmission parts (Chinchanikar and Choudhury, 2015) and dies (Dogra et al., 2010; Shibab et al., 2014). The advantage of hard turning over other processes is the reduction of equipment cost, flexible tooling (Bhemuni and Chalamalasetti, 2013), and complexity of the process due to the reduction of number of steps needed for machining a component (Dogra et al., 2010) as well as the ability of machining more complex

Computational Methods and Production Engineering. http://dx.doi.org/10.1016/B978-0-85709-481-0.00006-9

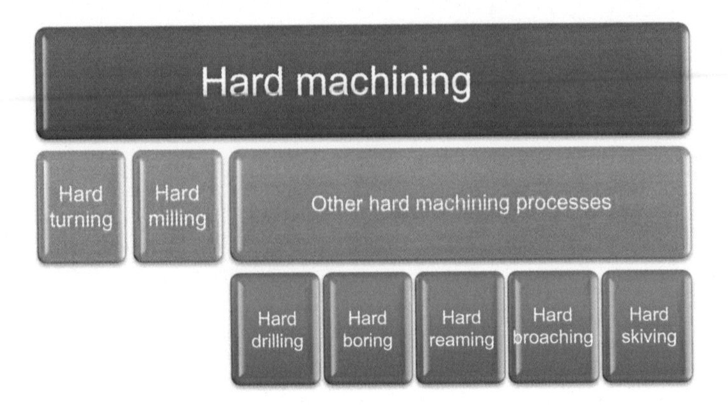

Fig. 6.1 Hard machining categories.

geometries and elimination of use of cutting fluid (Shibab et al., 2014). More specifically, material removal rates are higher in hard turning and machining setup time is less than in competitive processes such as grinding (Dogra et al., 2010). Thus hard turning can be viewed as a viable alternative of grinding operations (Astakhov, 2011); a brief comparison of hard turning and grinding operations can be found in the works of Grzesik (2008) and Toenshoff and Denkena (2013). Nevertheless, the range of cutting conditions which can lead to favorable outcome is somewhat narrower than in conventional turning and thus work is required to be done to determine this range of conditions for each case. Some of the characteristics of this process are high cutting speed, low feed, and low depth of cut (Bhemuni and Chalamalasetti, 2013), and the values of these parameters lie usually within the ranges: 55–400 m/min for cutting speed, 0.04–0.2 mm/rev for feed, and up to 0.2 mm or sometimes larger for depth of cut (Bartarya and Choudhury, 2012; Toenshoff and Denkena, 2013; Campos et al., 2014). A categorization of these conditions according to different cutting tools, derived from Campos et al. (2014), is presented in Table 6.1.

Astakhov (2011) commented that although a lot of research is still being conducted in the field of hard machining, only a few researchers realize that this machining category has significant differences from conventional machining processes and so they should be studied as a new category in many aspects.

Table 6.1 Typical process parameters for hard turning according to the cutting tool material

Cutting tool material	Cutting speed (m/min)	Feed rate (mm/rev)	Depth of cut (mm)
Ceramics	55–196	0.04–0.15	0.1–0.2
CBN	90–246	0.08–0.25	0.1–0.6
Cemented carbide	Up to 300	0.145	Up to 0.2

As for the machine tools and workpiece materials, hard turning exhibits in fact some particularities that need to be seriously considered. Workpiece is advised to have a small length-to-diameter ratio and complicated profiles are to be handled with care (Chinchanikar and Choudhury, 2015) in order to avoid chattering problems. Hard materials apart from the anticipated high hardness value, exhibit high abrasiveness and low ductility, properties that affect tool wear and formation of chip and cutting forces, respectively. Furthermore, the high value of hardness-to-elastic modulus ratio means that a considerable amount of local elastic recovery will take place after machining, affecting surface quality (Nakayama et al., 1988). Basic characteristics of the most common materials processed by hard machining are discussed in the work of Chinchanikar and Choudhury (2015). Machine tool requirements are high, as only machine tools with high rigidity are preferable for hard turning. It is advised that the whole machining system should be designed in a way that the rigidity is maximized and not only the machine tool alone (Astakhov, 2011). Moreover, although hard turning is a process in which no coolant is usually used, in cases that coolant is needed, mainly high pressure coolant is advised (Astakhov, 2011) as well as minimum quantity lubrication technique, which was shown to have beneficial effects (Grzesik, 2011).

From the relevant literature and practical application, it is considered that the use of appropriate tool material and type is essential for hard turning process, as it can prolong tool life and lead to optimal cutting conditions in terms of cost. However, in order for hard turning to possess advantages over the alternative processes, the cutting tools should meet several requirements. The requirements for cutting tools used for hard turning process are high hardness even at high temperatures, high compressive strength, high hardness-to-modulus ratio, high resistance to abrasion, thermal resistance, and adequate chemical stability at high temperatures, as material should withstand both high mechanical and thermal loads (Dogra et al., 2010; Shibab et al., 2014). Materials that exhibit high hardness such as CBN, PCBN, etc., are more often employed, as well other ceramic inserts, i.e., Al_2O_3 and Si_3N_4, or sintered carbides (Shibab et al., 2014), but it was shown that CBN tools can perform better than ceramics (Shibab et al., 2014). Without the use of the aforementioned types of cutting tools it would be impossible to machine workpieces at the 45–65 HRC hardness range (Dogra et al., 2010).

Another important factor concerning cutting tools in hard turning is the geometry of these tools, as high edge strength is required (Grzesik, 2011). As for the preferable tool geometry, similarly to other machining processes, tools with negative rake angle were found to be better performing in terms of tool life and larger nose radius was found to lead to better surface quality but generally increased cutting forces, whereas in the cutting edge zone the forces are kept within acceptable limits (Toenshoff and Denkena, 2013). However, smaller nose radius at high feed rates was observed to lead in larger white layer. According to the level of surface quality attained, hard turning can be categorized into rough, precision, and high precision hard turning (Grzesik, 2008). As for tool edge geometry, research has been already conducted and it was found that honed edges or the combination of chamfered and honed geometry is more preferable than sharp or chamfered ones; even tools with variable edge shape can lead to reduced wear and plastic deformation during hard turning as it provides higher strength to the CBN tool (Dogra et al., 2010). More specifically, chamfer is preferred between the rake and clearance faces and hones are often made additionally (Grzesik, 2011). Furthermore, tool coating during hard turning is advised, preferably with TiN-, TiAlN-, AlCrN-, or TiC-based coatings, with TiAlN being more preferable than others (Chinchanikar and Choudhury, 2015).

A significant amount of research has been conducted on determining the optimum process conditions that provide minimization of power, enhanced tool life, and better surface quality. The majority of these works are conducted with the aid of modern methods such as soft computing and statistical methods and representative works are mentioned and discussed in another section of this chapter.

As far as cutting forces are considered, it is generally anticipated that the high hardness values of the workpieces lead to a dramatic increase in cutting forces during hard turning (Shibab et al., 2014). Particularly in hard turning, due to the specific cutting conditions used, the thrust force is observed to be larger than in normal turning process and it also increases significantly as tool wear is increasing. This is mainly attributed to the large negative rake angle of the cutting tool and large corner radius (Grzesik, 2011). Due to high temperatures, usually exceeding 700°C in the chip-tool interface (Shibab et al., 2014) during hard turning, diffusion wear is more often occurring on tool material. As for chip morphology during hard turning, due to the reduced ductility of workpieces, sawtoothed chips or generally, severely segmented chips can be more often observed (Nakayama et al., 1988). The study on the mechanisms responsible for the transition from continuous to sawtooth chip type is still ongoing and more information can be also found in (Grzesik, 2011). Moreover, the high cutting forces are severely affecting the surface quality of the workpiece, as they often lead to the development of high tensile residual stresses, a completely not desirable situation due to the adverse effect on component fatigue life and other properties (Shibab et al., 2014). Consequently, a great part of research on hard turning is also concentrated on the determination of parameters that can lead to strong compressive residual stresses in the components and subsequently to a better surface quality.

A significant parameter regarding the rapid tool wear during hard turning is the high hardness of the workpiece material. This can be directly attributed to the microstructure characteristics of both workpiece and cutting tool. A secondary parameter affecting tool wear is the composition of tool and workpiece, which may favor tool wear due to chemical interaction. During these researches it was shown that mixed ceramic tools, such as Al_2O_3-TiC, may perform better, even from PCBN tools, in some cases (Grzesik, 2011). However, there does not exist yet a clear explanation for the increased tool wear during this process as more than one wear mechanism is often present. For deeper understanding of these subjects, the reader is advised to look at a thorough review of hard turning with CBN tools which was conducted by Dogra et al. (2010) who addresses the correlation between CBN properties with chip formation, cutting forces, and generally the outcome of the hard turning process.

6.1.2 Hard milling

Hard milling, as anticipated, refers to milling of hardened or tempered steels with high hardness values. It is the second most important hard machining process after hard turning and possesses many of its particularities, as well as possibilities for replacing other conventional and nonconventional processes. Hard milling is considered particularly appropriate for machining of machine tool components, for example, sliding guideway surfaces of machine tools (Toenshoff and Denkena, 2013). Another interesting example of use of hard milling is manufacturing of hollow tools intended for mold making and forming, such as plastic molds, forging dies, die casting dies. These components require partial hardening, as well as material removal after the hardening process. According to Grzesik (2008) hard milling has been proven as a viable alternative to other established nonconventional processes such as EDM in the mold making industry as it can create complex parts in one step, but lacks the ability of creating features that are more easily created by EDM, such as internal corners. Basic parameters of hard milling are the cutting speed, feed, axial, and radial depth of cut. According to literature on hard milling, recommended values for these parameters are 200–350 m/min for cutting speeds, 0.1–0.2 mm/tooth for feed (Toenshoff and Denkena, 2013), and depths of cut up to about 0.2 mm (Grzesik, 2008).

As for machine tools and material characteristics in hard milling, there exist also several particularities such as in hard turning. Some of the most widely used materials in the mold and die industry, such as P20, H13, W5, and S7 steels are usually treated by hard milling process. In order for the process to be conducted efficiently, some requirements should be met concerning the setup of the process. Machine tool should be capable of supporting processes like high-speed machining and the characteristics of the machine, e.g., the base of the machine should be designed in such a way that they ensure the reduction of vibrations. Finally, attention should be also paid to secondary elements of the setup such as toolholders as they should be able to function securely for a large range of rotational speeds (Astakhov, 2011).

Cutting tools for hard milling belong usually to the following three types: solid carbide end mills, indexable carbide inserts, and ceramic indexable inserts. As far as solid carbide end mills are concerned, these tools can provide better surface accuracy but they are the most expensive tools. Astakhov (2011) commented that carbide inserts are considered to be less reliable as they suffer from reduced tool life due to the fact that usually they are not adjusted to the particularities of hard milling. Finally, whisker-reinforced ceramics are a very promising category of tools for hard milling as they offer various advantages: faster production times, security at high velocities, superior performance than carbide inserts at high temperatures. Some of the main issues related with the tool wear occurring in hard milling process are the abrasive flank wear on the insert and frittering, i.e., creation of multiple cutting edges of the workpiece.

During hard milling it is recommended that positive insert geometries with sharp edges are used in order to reduce cutting forces. Furthermore, it is advised that use of cutting fluid should be avoided, mainly when ceramic inserts are used and trochoidal milling should be considered to enable high table feeds and low cutting forces (Astakhov, 2011). In fact, in hard milling, higher cutting speeds and feed rates are noted to be used, compared to hard turning or drilling due to the fact that in hard milling the cutting edge engages periodically with the workpiece and not during the whole process, as in hard turning. This results to a sudden shock in terms of thermal and mechanical load but in average this loading is lower than in other processes (Toenshoff and Denkena, 2013).

6.1.3 Other hard machining processes

Despite the fact that hard turning and hard milling are the most important hard machining processes in practice, in the relevant literature exist several other machining processes that can be performed to create components from hardened steels. One of the most important is hard drilling. This process is essential to the manufacturing of holes, regardless of whether another process had been previously performed. Compared to other hard machining processes, process parameters in hard drilling are required to lie in a narrower range, with feed in the range of 0.01–0.063 mm and cutting speeds up to 160 m/min (Toenshoff and Denkena, 2013). Preferable cutting tools for hard drilling are solid carbide drills, while drilling tools with inserts and TiN coating are used for large hole diameters.

Some less frequently used machining processes are (Grzesik, 2008; Astakhov, 2011) hard boring, hard reaming, hard broaching, and hard skiving.

6.1.4 Phenomena occurring during hard machining processes

Some important phenomena related to hard machining processes are the microstructure-related phenomena of white and dark layer formation, the development of residual stresses beneath the surface, and generally the surface quality of the machined component. Two general theories were proposed to explain the main phenomena occurring during hard turning with the most accepted being that a high temperature zone is developing in front of the cutting tool during the processes, leading to material softening in this area, as a process similar to annealing occurs, and thus enabling the cutting process to be more easily conducted (Bartarya and Choudhury, 2012). However, another proposed theory states that plastic deformation through high compressive stresses is the dominant factor in this process. These phenomena related directly to the machining conditions lead to the final structure and properties of the workpiece, a topic that is still intriguing for researchers.

As for material microstructure after hard machining, a hard-machined surface is structured in three layers, namely, the white layer consisting of untempered martensite, the dark layer consisting of overtempered martensite, and finally the bulk material. Each layer has its own structure and the thickness of each layer varies with the machining conditions. White and dark layers are associated with tensile stresses beneath the surface; they reduce considerably the fatigue life of parts (Campos et al., 2013).

More specifically, the white layer is a modified martensite structure which appears to be white after etching when viewed through an optical microscope (Campos et al., 2014). This untampered martensitic structure has higher hardness than the hardness of the bulk material and dark layer. Theories on the formation of white layer have been developed according to which the white layer is the result of the thermal process during which a phase transformation occurs due to severe plastic strain, possibly caused by the increased workpiece-tool friction (Campos et al., 2013). However, other mechanisms contributing in the formation of white layer can also be the rapid heating and quenching of the workpiece or surface reactions with the environment. TEM analysis of white layer has indicated that two different mechanisms are responsible for the formation of this layer regarding the cutting speed at which the process was conducted, with severe plastic deformation being the cause at low-to-moderate cutting speeds and thermally driven phase transformation being the cause at high speeds (Bartarya and Choudhury, 2012). The microhardness values within the white layer are reported to be considerably higher than in the other layers of the material and a sudden decrease in hardness is observed within the dark layer due to softening. Grzesik (2008) stated that the thermally affected zone is typically the 1/10 of the depth of cut and that the thickness of white layer increases considerably with increasing tool wear and cutting speed up to a point. Both white and dark layers are brittle and prone to cracking so they are considered as damaged part of the workpiece.

6.2 Numerical modeling of hard machining

Although the use of a numerical method for predicting the outcome of the machining process is not new, the use of models specialized to the machining of hard-to-cut materials is relatively more recent. As it will be discussed afterward in more detail, these advanced models include special treatment of process conditions such as boundary conditions or material models, which reflect the particularities of hard machining.

The general models of orthogonal or three-dimensional machining or thermal models are adapted with a view to study thermal and thermomechanical phenomena within the workpiece, tool wear phenomenon, and microstructural alterations of the workpiece materials such as the formation of white and dark layers. During the last years, the works in the relevant literature are increasing and some of the most representative are summarized later.

6.2.1 Thermal models

As it was mentioned before, one of the modeling approaches concerning hard machining is one of the thermal phenomena occurring during these processes. After the necessary assumptions are made, the heat produced by the friction of cutting tool-workpiece interface is replaced by a heat source and partition of heat between the components of the process is calculated. Although these models are somewhat less complex than the others, the determination of temperature field and temperature gradients can provide useful information about heating and cooling rate, heat-affected zone, formation of white and dark layers, and other microstructural alterations; for models predicting hardness or microstructure, more details are given in the following section. Afterward, some characteristic works regarding thermal models of hard machining processes are presented.

Sukaylo et al. (2004) created a numerical model to predict workpiece distortions due to thermal effects of hard turning. In order to develop the numerical model based on experimental data, they first conducted experiments of dry hard turning of 20MoCrS4 steel with a depth of cut of 0.3 mm, feed rate of 0.1 mm/rev, and values of cutting speed in the range of 90–180 m/min, whereas tool wear was also taken into account. Heat transfer coefficients were determined by a previously developed model. By the validation of the numerical model, it was found that due to its high conductivity, the CBN tool removes a considerable amount of heat of the workpiece, resulting in rather low temperature increase, and also the effect of process parameters to temperature field and generated heat was also determined. Finally, this model was applied to calculate workpiece distortion due to heat phenomena. Kryzhanivskyy et al. (2015) developed a model to predict cutting temperature during hard turning with PCBN cutting tools. Their model was also based on experimental observations of temperature at various locations of the cutting tool. Several sensors were placed at the cutting tool during the experiments to record temperature values, while cutting forces and tool wear were also measured. A two-dimensional model was created to solve the inverse thermal problem of determining the values of heat transfer coefficients between the various parts of the cutting tool (subdomains) as well as the respective heat fluxes. Material properties for the various subdomains were obtained from the literature and measurement of total power, cutting forces, and tool wear were also used. From the solution of the inverse problem, the partition of heat fluxes between the various subdomains of the cutting tool was able to be determined as well as heat transfer coefficient, with a significant level of accuracy, for various cases of tool wear and machining time. Kundrák et al. (2017) developed a finite volumes thermal model to investigate various aspects of the thermal phenomena during hard turning such as heating and cooling rate, white layer formation, etc.

6.2.2 Thermomechanical models

Although the thermal models can provide sufficient preliminary results concerning thermal phenomena in the workpiece, it is more usual to adapt a classical orthogonal cutting model to model hard machining and predict the combined thermomechanical phenomena occurring during this process, something that results in more demanding models but also in a clearer understanding of the process. In this section, several thermomechanical models are presented and discussed. In the following section, more advanced models in which the effect of initial material hardness and other additional elements to the flow stress are also taken in consideration will be presented.

Ng and Aspinwall (2002) conducted a Finite Elements Method (FEM) study of hard machining with emphasis on chip morphology. They used Johnson-Cook constitutive model with failure criterion for the workpiece material and two different models, one with crack nucleation and one without which resulted to prediction of serrated and continuous chip, respectively. The results concerning shear angle, temperature, and stress distribution were evaluated and it was found that the experimental results were closer to the model with serrated chip and in general trends concerning the machining parameters and the outcome of the process were accurate.

Guo and Liu (2002) developed a FEM model with a thermo-elastic–plastic material model and improved friction coefficient model. Using this model they were able to calculate cutting forces and residual stresses. Investigation on the sensitivity of material failure strain, material flow stress, and friction coefficient was also conducted. Guo and Yen (2004) studied the mechanisms of discontinuous chip formation during hard machining with the aid of a FEM model. A two-dimensional plain strain model with Johnson-Cook plasticity model and Johnson-Cook failure model was created to predict chip morphology at high cutting speeds (1000 m/min) and special care was given to the adaptive meshing, as it was expected that significant distortion of elements could occur. This model enabled the study of chip formation during all the stages of the machining process, as well as the cutting forces progression without the use of specially created subroutines.

Cakir and Isik (2005) created a structural model of a cutting tool in order to predict tool failure during hard turning. For this study experimental data for cutting forces during hard turning were gathered using a tool breakage monitoring system. Then, they conducted a static FEM analysis using the cutting force values obtained from the experiment, in order to determine the stress distribution in the cutting tool, before breakage occurs. The next step was modal analysis, during which natural frequencies and modes of the system are determined and the final step was harmonic analysis with a view to fully determine the dynamic behavior of a hard turning cutting tool regarding failure.

Arrazola and Ozel (2008) created an Arbitrary Lagrangian-Eulerian (ALE) model for the simulation of three-dimensional hard turning in order to investigate the effect of tool edge geometry on cutting forces, stresses, and temperatures. In order to produce a realistic chip that is difficult when using complex tool geometries, and ALE method with Eulerian boundaries, they proposed an additional remeshing step to be added. In total, three intermediate remeshing steps between the initial and the steady-state step were performed. Thus the mesh was sufficiently accurate for the various stages of the process, artificial criteria for chip formation are not required and accuracy of results is higher. Mamalis et al. (2008) presented a paper with orthogonal and oblique cutting FEM models of high-speed hard turning. The combination of high speed with hard turning is quite advantageous since better final product quality, reduced machining time, lower cost, and environmentally friendly characteristics can be achieved. For the FEM modeling, Third Wave Systems Advant-Edge software was used and the obtained results contained forces, stresses, and temperatures within the workpiece and the tool (see Fig. 6.2).

Mankova et al. (2011a) carried out simulations and investigated the effect of cutting parameters during hard turning. More specifically, they created a model for hard turning of AISI 1045 by mixed oxide ceramics for five values of cutting speed in the range of 90–350 m/min and four values of feed in the range 0.047–0.2 mm/rev. The model was compared to experimental results and it was found that a difference of

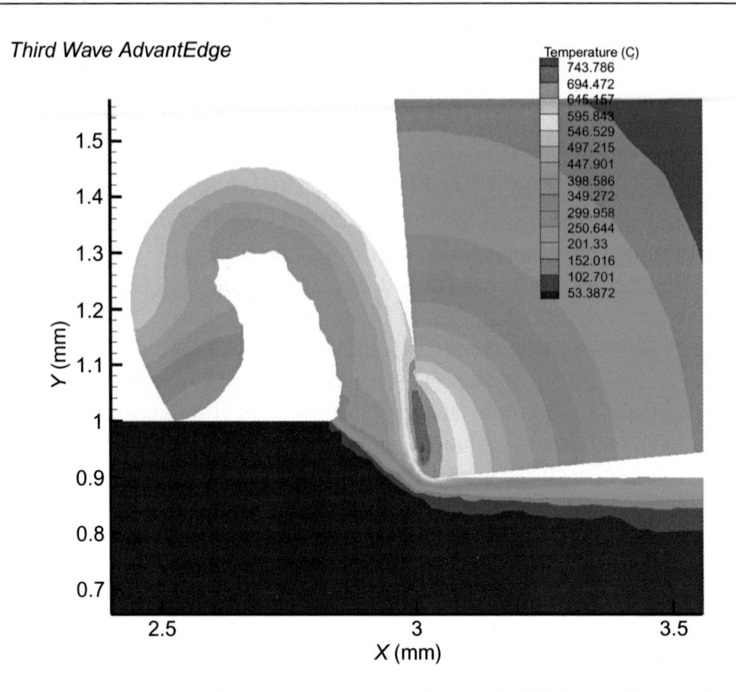

Fig. 6.2 Temperature contours in the workpiece and the tool in FEM modeling of high-speed hard turning.

less than 10% was obtained in all cases. Using this model, the effect of cutting parameters on cutting forces, deformation of workpiece, chip formation, and temperature field and heat fluxes on workpiece and cutting tool were determined. Mankova et al. (2011b) carried out further experiments with the same cutting parameters' values but they considered the effect of material hardness of the workpiece as well.

Bagci (2011) conducted a study of hard turning using a Smoothed Particle Hydrodynamics (SPH) model in order to find out if it could perform better than ALE methods in the case of large distortions. A plastic kinematic hardening model and the Cowper-Symonds model for strain rate was used in the SPH model and comparison between SPH, experimental results, and results with other methods was done. In both simulation cases, results exhibited a deviation of 8%–11% and in the case of normal cutting force, it performed better than FEM model. It was proposed that more detailed investigation of SPH application to machining cases should be done in order to increase accuracy of models. Szabó and Kundrák (2012) conducted two-dimensional FEM simulations of hard machining at various conditions with emphasis on chip morphology and especially the so-called sawtooth formation.

Takacs and Farkas (2014) conducted simulations of hard cutting of AISI D2 steel with a FEM model under various conditions. In specific, they carried out experiments for five values of feed rate (0.03–0.2 mm/rev) and five values of cutting speed (100–250 m/min) and measured the three components of cutting forces. In the FEM model, a geometrically accurate model of the cutting insert was used and simulations with five values of feed rate were conducted. The relatively high differences of the produced results from the experimental values were attributed to the constitutive model applied or the fracture criteria and they also noted the need of additional remeshing at specific zones of the model.

Bapat et al. (2015) created a two-dimensional numerical model to study temperature distribution during turning of AISI 52100 steel. Johnson-Cook constitutive model as well as Johnson-Cook failure model was selected for the workpiece material and thermal expansion coefficient. Young's modulus and Poisson ratio were considered temperature dependent. Temperature distribution and chip morphology were obtained and the results were compared to experimental ones, showing an average error of 13%. Kim et al. (2015) conducted a novel study of simulating hard turning process with a micro-textured tool. The cutting tool was modeled as a deformable body and four in total different micro-textured geometries were considered, namely, nontexture, perpendicular, parallel, and rectangle. A Johnson-Cook material model was applied and properties such as elastic modulus, Poisson ratio, thermal conductivity, expansion coefficient, and heat capacity were modeled as temperature dependent; friction values were different for each case. After comparisons with experimental results were conducted, the effect of various micro-texture characteristics as well as friction on cutting forces was investigated.

Kundrák et al. (2016) investigated an important aspect of surface integrity, namely, residual stresses. These stresses are induced in the workpiece during machining and tool rake angle plays an important role on their features, regarding the stresses' magnitude and distribution within the workpiece. In this chapter, numerical investigations with the use of FEM were presented that allowed the evaluation of the influence of the tool rake angle, in a wide range of positive and negative rake angles, on residual stresses for the case of hard turning of stainless steel. Klocke et al. (2016) employed an innovative method in order to correlate empirically obtained characteristics with calculated loads. More specifically, they first conducted hard turning experiments with three different cutting conditions in order to determine the effect of cutting speed and feed on workpiece surface alterations. From these experiments, tool wear and dimensions of the heat-affected zone and white layer were observed. Afterward, a three-dimensional FEM model was developed with a view to calculate temperatures and stresses during hard turning process for the cases which were studied experimentally. Finally, the method was applied successfully to compute thermomechanical loads at the interfaces between bulk material, heat-affected zone, and white layer zone and to attempt to explain their formation.

6.2.3 Thermomechanical models with flow stress incorporating initial hardness

Whereas the simpler thermomechanical models can provide the researchers with a fairly good understanding of the effects of input conditions to the outcome of the process and be particularly useful for several cases, researchers have also developed specialized material models for the case of hard machining, taking into consideration the effect of initial hardness and other elements which can describe the material behavior more accurately. Despite the fact that these models can perform significantly good in specific cases, research is still conducted in this field, as there does not exist yet a globally accepted model that can perform sufficiently in every case and many modifications can be proposed for the creation of more realistic models.

Poulachon et al. (2001) were among the first researchers who discussed the particularities of hard turning modeling and demonstrated the necessity of developing suitable flow stress laws to explain the material behavior during hard turning. Yan et al. (2005) conducted simulations of finish hard turning, with the aid of a FEM model. For material modeling, they employed a previously derived flow stress law given by the following formula:

$$\overline{\sigma}\left(\overline{\varepsilon},\dot{\overline{\varepsilon}}T, HRC\right) = (A + B\overline{\varepsilon}^n + C\ \ln(\varepsilon_0 + \varepsilon) + D)\left(1 + E\ \ln \dot{\overline{\varepsilon}}^*\right)(1 - (T^*)^m) \tag{6.1}$$

where ε is the strain; $\dot{\overline{\varepsilon}}^*$ represents the dimensionless strain rate as the ratio of $\dot{\overline{\varepsilon}}$ to $\dot{\overline{\varepsilon}}_0$ ($\dot{\overline{\varepsilon}}_0$ is equal to 1.0 s^{-1}); ε_0 is the reference strain (assumed to be 10^{-3}); T is the workpiece temperature in Kelvin; and T^* is the homologous temperature, taking into account the reference temperature and melting temperature. A, B, E, n, and m are constant material properties for AISI H13 steel and finally C and D are parameters related to initial material hardness. In this formula, hardness of the workpiece is expressed in terms of Rockwell-C scale hardness. The expressions for C and D are given as follows:

$$C(HRC) = 0.0576 \times (HRC)^2 - 3.7861 \times HRC + 52.82(\text{MPa}) \tag{6.2}$$

$$D(HRC) = 0.6311 \times (HRC)^2 - 12.752 \times HRC - 727.5(\text{MPa}) \tag{6.3}$$

This model enabled an accurate prediction of chip morphology and the trends between machining parameters, forces, and temperatures during hard turning.

In their work, Umbrello et al. (2004) stated that material models for hard machining should account for deformation response to high loading rates, large deformations, and large changes in temperatures and reported several models capable of capturing complex behaviors such as Johnson-Cook, Zerilli-Armstrong (Zerilli and Armstrong, 1987), Steinberg (Steinberg et al., 1980), and Goldthorpe-Church (Goldthorpe and Church, 1997), some of which can include the effect of material composition, dislocations, grain size, phase changes, etc. The flow stress law can be described by the following formula:

$$\sigma(\varepsilon,\dot{\varepsilon},T, HRC) = B(T)(C\varepsilon^n + F + G\varepsilon)[1 +\ \ln(\dot{\varepsilon})^m - A] \tag{6.4}$$

This formula is comprised of four parts, each related to a specific material behavior: the $C\varepsilon^n$ term is related to *work hardening*, $B(T)$ is the influence of temperature (temperature softening), $1 +\ \ln(\dot{\varepsilon})^m - A$ is related to the influence of strain rate, and $F + G\varepsilon$ term represents the influence of hardness. More specifically, $B(T)$ is an exponential function of workpiece temperature produced by experimental tests. The formula for determining $B(T)$ is as follows:

$$B(T) = \exp\left(aT^5 + bT^4 + cT^3 + dT^2 + eT + f\right) \tag{6.5}$$

where a, b, c, d, e, and f are material constants and 20°C is the reference temperature. The data for the work-hardening term were derived from the literature for AISI 52100 steel, as well as the data for the strain rate term. For the derivation of material hardness (HRC) effect term, a special procedure was carried out. This procedure was applied to data available from the literature for a wide range of hardness values and using regression method they calculated F and G parameters, which are given from the following formulas as linear functions of hardness:

$$F(HRC) = 27.4 * HRC - 1700.2 \tag{6.6}$$

$$G(HRC) = 4.48 * HRC - 279.9 \tag{6.7}$$

From the previously mentioned two formulas, it can be shown that parameter F is modifying the initial yield stress value and G the strain hardening curve. Brozzo's criterion in order to include the effect of hydrostatic stress on the chip segmentation was also used. Afterward, in order to validate their model, they conducted experiments and compared the cutting force values and chip morphology with the predicted one for four different cases, observing a very close match between the two.

Hua et al. (2006) investigated the optimum cutting conditions and cutting edge preparation regarding subsurface residual stresses during hard turning. In order to include material hardness in their simulations, they employed the flow stress law originally introduced by Umbrello et al. (2004), which has some similarities to the material law described in Yan et al. (2005). The numerical model succeeded particularly in predicting the residual stresses distribution from a depth under the surface, whereas deviations were observed in this profile near the surface. Umbrello et al. (2008) developed later a more advanced model for the flow stress during hard turning using a specific iterative procedure for calculating the effect of initial hardness on reference stress curve, which in their study was the Johnson-Cook model proposed by Shatla et al. (2001). The iterative procedure was applied to data from the literature and regression analysis was employed to derive two third-order polynomial functions for F and G material parameters of their model. This model was calibrated with experimental data for various cases of hard turning and then validated using experiments. It was shown that this model had an average accuracy of 10%–15%.

Kountanya et al. (2009) used both the flow stress model proposed by Umbrello et al. (2004) and Huang and Liang (2003) in their studies of hard turning. They observed that each model was particularly capable of representing a particular range of strain, strain rate, and temperature values and thus a combination of the two models was conducted. After other modifications, such as the calculation of contact length between tool and chip by experimental observation, were added, they employed this model to investigate the effect of catastrophic thermoplastic instability and surface shear-cracking mechanisms to chip formation and machining forces.

Uhlmann et al. (2015) created a new material model for hard turning processes, which can consider the viscoplastic asymmetry effect. More specifically, they included the transformation-induced plasticity (TRIP) by the use of weighting functions under various stress conditions. The Leblond approach was also considered to modify the Johnson-Cook model for the ductility alteration caused by TRIP. Experiments with variable cutting speed, feed, and depth of cut were conducted and then the material models were developed using the aforementioned approaches. With this model, the phase transitions between martensite and austenite, the formation of white layer, and the correlation of austenite start temperature with the externally applied stress were studied.

Jivishov and Rzayev (2016) compared various material models used in FEM models for the case of hard turning. Initially, the models which they compared were Johnson-Cook, Koppka (Koppka et al., 2001), El-Magd (El-Magd and Treppmann, 2000), and Oxley (Oxley, 1989) models and necessary parameters for each models were obtained for AISI 1045. Then, they compared Poulachon (Poulachon et al., 2001), Poulachon-IEP (with the same equation as Poulachon but with experimental constants obtained by Institut of Experimental Physics, University of Magdeburg), Huang (Huang and Liang, 2003), and Umbrello (Umbrello et al., 2004) models for the case of AISI 52100 steel. The comparison involved the prediction of stresses and cutting forces for cases with various depths of cut and they found that for AISI 1045 Oxley model is better for cutting forces prediction whereas Poulachon's material model is better for AISI 52100 steel.

6.2.4 Thermomechanical models with microstructure-related prediction capabilities

Although it is already mentioned that several thermomechanical models which were derived for hard machining can predict the outcome of this process in a realistic way, there exists also a promising and interesting subcategory of thermomechanical models, in which microstructure-related phenomena can be modeled and analyzed. With the aid of advanced experimental techniques, results concerning quantities such as grain size can be incorporated into the models by special formulas and lead to the prediction of microstructural alterations within the workpiece, as well as improve the prediction of other more macroscopic phenomena. As the majority of these models are relatively new, it is expected that in the future it will be an increase in the interest of researchers to develop models pertinent to this category.

Caruso et al. (2011) developed a FEM model to predict microstructural changes in hard turning. They put an emphasis on the alterations of grain size during hard turning, something that should be incorporated in the thermomechanical model if an accurate simulation of white and dark layer is required. Initially, they conducted experiments for hard turning cases with three different cutting speed values after which they determined grain size. Afterward, they used the Zener-Hollomon relationship in order to include the effect of microstructure changes in the model by the following procedure: they compared the actual strain to the critical value and recrystallization was occurring only if the actual value exceeded the critical one; after recrystallization was finished, a new value for grain size was computed. However, further calculations were needed to determine parameters of the model, based on experimental data from the literature. Then, results concerning grain size and thickness of white and dark layers were obtained from the model and compared to the results of the three experiments exhibiting good levels of agreement.

Ding and Shin (2013) presented a multiphysics model with a view to predict surface microstructural alterations during hard turning. Their study was among the first to take into consideration phase transformation and grain refinement and the model was capable of solving simultaneously for heat transfer, plastic deformation, and microstructural alterations. For the validation of the model, experimental results from the literature were also used. Johnson-Cook model was selected for the workpiece material under austenitic temperature and a special model by Iwamoto was selected for austenite and martensite phases. In order to model the phase composition of the workpiece during each stage of the process, phase transformation kinematics were solved at each step. Austenitization temperature was defined, special models were used to predict phase fractions, and transformation plasticity was incorporated into the model. Furthermore, grain evolution and dislocation density were also calculated at each step. Results concerning temperature, phase composition, microhardness, and white layer thickness were presented, with significant level of accuracy compared to experimental data of residual stresses, white layer thickness, and microhardness and explanations on the nature of each phenomenon were derived.

Rana et al. (2016) created a FEM model for the prediction of surface alteration during hard machining of Inconel 718 at various conditions. Before the model was created, experiments at three different cutting speeds under dry cutting conditions were carried out and after each experiment, cutting forces, chip morphology, and microhardness of workpiece were obtained. For the numerical model, Cockroft and Latham's damage criterion was used for chip segmentation prediction, a hybrid friction model was selected, and special model for grain size calculation was used along with Zener-Hollomon relationship and Hall-Petch equation for hardness variation predictions. This model was found to be able to predict cutting forces, grain size, and microhardness with less than 8%, 7%, and 2% deviation, respectively. Fergani et al. (2016) created a FEM model to predict microstructure texture using Visco-Plastic Self-Consistent (VPSC) methodology. Experimental tests with two different rake angle values and two different cutting speed values were conducted. For the numerical model, Johnson-Cook constitutive model and Johnson-Cook failure model were selected and special subroutines were created in order to calculate deformation gradient and velocity gradient. It was found that this model could predict body-centered cubic (bcc) texture with an acceptable level of accuracy in comparison to experimental results. Further information on advanced modeling of hard turning regarding microstructure can be found in the work of Umbrello et al. (2010).

6.3 Soft computing and statistical methods modeling of hard machining

Apart from traditional simulation methods, simulations of hard machining processes using soft computing and artificial intelligence methods have been also conducted. These methods derived initially from the computer science and robotic-related research but during the last decades they have been employed in many engineering disciplines with a view to predict complex phenomena for which no other analytical solution exists or the solutions are very time consuming. Fig. 6.3 presents a general outline of the most commonly used soft computing and statistical methods categories. Although these methods do not model explicitly the physics of the process, they are still useful to determine the correlation between input and output parameters when these are highly nonlinear. Apart from the methods presented in this section, recently there was a research work by Agrawal et al. (2015) in which two promising methods, namely, random tree forest regression and quantile regression, were employed. For a more comprehensive reading on soft computing and statistical methods, the interested reader can search the work by Markopoulos et al. (2016).

6.3.1 *Artificial neural networks*

Artificial neural networks (ANN) are using the principles of biological neural network function to predict the correlation between a set of input and output parameters. These networks can be used both in classification, e.g., assigning individual items to

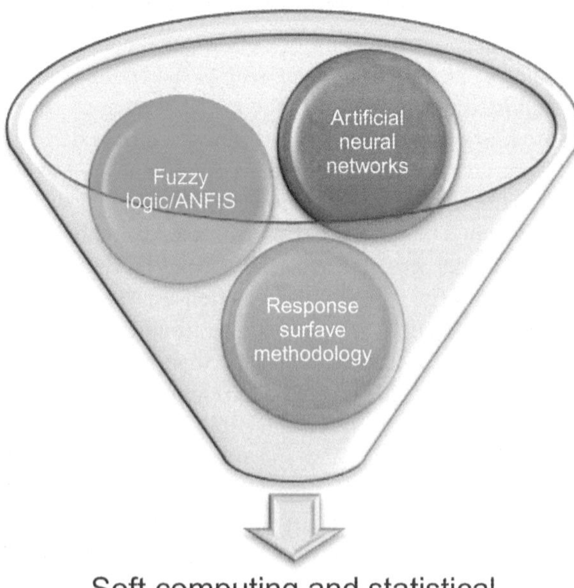

Fig. 6.3 Soft computing and statistical methods categories often used in hard machining modeling.

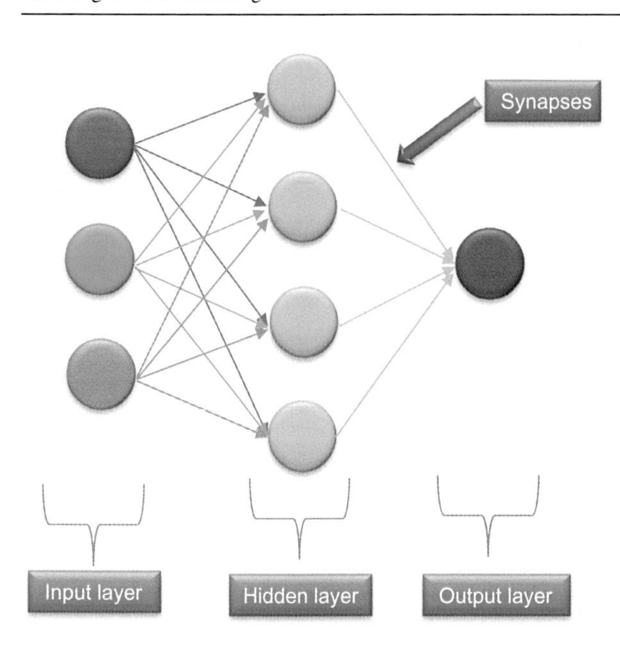

Fig. 6.4 Example of the architecture of a back-propagation feed-forward ANN.

categories according to their value and regression problems, where the correlation between input and output is required. The network architecture for a usual back-propagation feed-forward network is presented in Fig. 6.4 that depicts the basic components of the network: neurons arranged in groups or layers interconnected by synapses, each of them having a weight coefficient. The network is fed with series of inputs and outputs and after periods of training where weights are adjusted according to error, the network is able to reproduce reasonably predicted outputs, given a set of suitable inputs. The quality is evaluated according to mean square error of the prediction or coefficient of determination between actual and predicted output. Some representative works on hard machining modeling with ANN method are described afterward.

Özel and Karpat (2005) developed predictive models for the estimation of tool wear and surface roughness during hard turning using ANN models. The model was created to predict both flank wear and surface roughness and had five inputs, i.e., edge geometry, hardness, cutting speed, feed rate, and cutting length. Furthermore, four different neural network models with seven inputs and one output were created. The inputs were workpiece hardness, cutting speed, feed rate, axial cutting length, and mean values of the three force components measured during finish hard turning. It was found that ANN models were better than regression models and the ANN model with forces as input and a single output performed better than the other ANN model.

Sharma et al. (2008) created a model to predict both cutting forces and surface roughness during hard turning. Input parameters were approaching angle, cutting speed, feed rate, and depth of cut. It is to be noted that the authors determined optimum number of neurons and epochs for the training using regression equations and optimization process before creating the final neural networks models. The developed models were able to indicate the correlations of input variables to the outputs.

Quiza et al. (2008) conducted experiments on tool wear during hard turning and developed statistical and neural networks models for the prediction of flank wear. The parameters for Multilayer Perceptron (MLP) model were selected after an investigation using Taguchi method and Analysis of Variance (ANOVA) took place. The optimum neural networks model was found to be significantly superior to the statistical one.

Wang and Yuang (2009) employed a Recurrent Neural Network (RNN) model for the estimation of tool wear in hard turning. The experiments were conducted for AISI 52100 bearing steel with four variable parameters, i.e., cutting speed, feed, depth of cut, and machining time. The number of hidden layer neurons for the RNN and optimized RNN model varied from 1 to 12 and finally, the architecture 5-2-1 was selected. In comparison with other neural network types, the RNN and optimized RNN networks performed generally better with a considerable reduction of error in some cases.

Al Hazza and Adesta (2011) developed a model for flank wear estimation during high-speed hard turning. They included three parameters as inputs at four levels and conducted experiments using a full factorial design. The experiments were conducted at dry cutting conditions with a constant depth of cut of 0.15 mm. In total the developed ANN model had three input layer neurons, i.e., cutting speed, feed rate, and rake angle; 20 hidden layer neurons; and 1 output neuron. This model was slightly better than a regression model and had a deviation of 4% from the experimental results. Abdullah et al. (2012) employed an ANN model for the estimation and optimization of cutting conditions in respect to surface roughness during hard turning. They conducted experiments in which the three input variables were cutting speed, feed, and depth of cut and these experiments were designed using a L_{27} orthogonal array. The ANN model had architecture of 3-23-9-1 and Mean Squared Errors (MSEs) were below 0.004, slightly better than MSEs of a regression model.

Beatrice et al. (2014) used ANN models to study surface roughness during hard turning with minimal cutting fluid. Input parameters were feed rate, cutting speed, depth of cut and output parameter was surface roughness, whereas the experiments were conducted using L_{27} orthogonal array. Several network architectures were tested, according to their RMS error and 3-7-7-1 network was found to be the best, as it exhibited sufficient accuracy (error $<7\%$) and R value of 0.96. Arulraj et al. (2014) created a sensor fusion model based on neural networks for surface roughness prediction during minimal cutting fluid conditions. Input parameters were feed rate, cutting speed, and depth of cut and 27 experiments were carried out, according to Taguchi L_{27} array. Models with and without sensor fusion were compared and it was found that models with sensor fusion were significantly superior to those without fusion, both regarding error of prediction and coefficient of determination.

Senthilkumar and Tamizharasan (2015) created an ANN model for the prediction of flank wear of the cutting tool and surface roughness of workpiece during hard turning. They selected seven input parameters for the model, namely, cutting speed, feed rate, depth of cut, material type, cutting insert shape, relief angle, and nose radius of the cutting tool and conducted experiments using a L_{18} Taguchi orthogonal array. Outputs of the neural networks were flank wear and surface roughness. ANN results were sufficiently accurate in comparison to those of a regression model and this accuracy was further justified by confirmation experiments.

Finally, Mia and Dhar (2016a) employed an ANN model in order to study the development of surface roughness profile during hard turning with High Pressure Coolant (HPC). In their work, they conducted the experiments using a full factorial design of four input variables, i.e., cutting condition, cutting speed, feed rate, and workpiece hardness at various levels, for a total of 96 experimental runs. As cutting condition was a categorical variable (dry or HPC conditions) they decided to develop both separate networks for dry and HPC turning and a network with cutting condition as input. Three different training algorithms were compared, namely, Levenberg-Marquardt, Bayesian Regularization, and Scaled Conjugate Gradient. Optimum architectures and training algorithm were determined for the two types of neural networks but finally the network with three input parameters was the better as it exhibited the lower root-mean-square error (RMSE).

6.3.2 Fuzzy logic models

Although there is not yet a large amount of works conducted concerning modeling of hard machining with fuzzy logic and adaptive network-based fuzzy inference system (ANFIS) methods, it is worth discussing some of the existing works in the literature.

A very useful application of fuzzy logic models concerning hard machining is the determination of optimum process parameters or even tool geometry given the characteristics of workpiece and tool material. Hashmi et al. (1998) created a fuzzy logic model for the selection of optimal cutting conditions in machining processes. The model was created using data from machining experiments with two types of steel, three different depths of cut, and four types of cutting tools and material hardness and cutting speed were chosen as input and output variables, respectively. The model included six rules and triangular membership functions and it was noted that predicted values had a very good correlation with values recommended by machining handbooks. Wong et al. (1999) used a fuzzy logic model with depth of cut as an additional input to material hardness and used two separate outputs, namely, cutting speed and feed rate. For the cutting speed fuzzy model, five levels were defined for material hardness, as well as for depth of cut, whereas fifteen levels were defined for the cutting speed; triangular shape membership function was selected for every input and output. For the feed rate model, five levels were defined for depth of cut and five for feed rate. Fuzzy rules were applied in a straightforward way to the models. Finally, separate models were created for different types of tool materials, namely, high-speed steel tool, uncoated brazed carbide tool, uncoated indexable carbide tool, and coated carbide tool. Errors were found to be at acceptable levels and results of the fuzzy logic models were considered adequately reliable in order to be included in a process planning system.

Osorio et al. (2007) employed fuzzy logic models in order to create a system, which was able to suggest the appropriate cutting tool geometry for several work-tool material combinations. They used several workpiece material as well as hardened steel (AISI 4340) to create the fuzzy logic model. Inputs to this model were the work material specific cutting energy and the "destruction" specific energy of the tool material and outputs were cutting tool characteristics. Due to the fact that available data did not cover adequately all the range of input parameters, two separate fuzzy logic models were designed for each output parameter—six in total—and the results were evaluated by using a reliability percentage index.

Akkus and Asilturk (2011) used a fuzzy logic model along with ANN and regression models in order to compare their efficiency toward modeling of surface roughness during hard turning. Workpiece material was AISI 4140 and input parameters for the models were cutting speed, feed, and depth of cut. For the model, Mamdani approach was adopted, the triangle membership function was selected, inputs were divided into three linguistic expressions, and outputs into eleven linguistic expressions, while 27 fuzzy rules were established in total.

Tamilarasan et al. (2015) used a fuzzy logic model for the simulation of hard milling. Experiments on 100MnCrW4 (Type O1) steel material were conducted, designed using a Central Composite Rotatable Design (CCRD) method, and inputs to the fuzzy logic model were feed per tooth, radial depth of cut, axial depth of cut, and cutting speed, whereas outputs of the model were cutting temperature, tool wear, and material removal rate. Triangular membership function was employed and thirty fuzzy rules were constructed. Results using this model were evaluated according to R^2 and average error values and compared to a regression model. Although fuzzy logic model performed slightly worse than the Response Surface Methodology (RSM) model, the results indicated that the model was sufficiently accurate. Zuperl and Cus (2015) included a surface roughness ANFIS predictive tool in an adaptive force regulation system for end milling. For the ANFIS model they employed a dataset of 350 sets of experimental data of hardened die steel machining and they used force sensor readings, feed rate, spindle speed, radial depth of cut, and axial depth of cut as inputs, while surface roughness was the output. Triangular membership function was selected and 36 rules were used in the model in total and high accuracy of the predicted results was noted. Saini (2015) implemented a neuro-fuzzy model to predict residual stresses, surface roughness, and tool wear during hard turning. He conducted 29 experiments on AISI H11 workpiece based on Box-Behnken approach. Inputs of the ANFIS system were cutting speed, feed, depth of cut, and nose radius, while the outputs were axial residual stress, surface roughness, and tool wear. Separate models were created for each output, as it was supposed that there is no correlation between them. Results of the ANFIS model were compared to those of a regression model and their efficiency was evaluated using a Chi-square (χ^2) test for the goodness of fit. The ANFIS model was found to perform better for residual stresses but was outperformed by the regression model in the case of surface roughness and tool wear prediction.

Cica et al. (2015) conducted a comparative study concerning soft computing techniques in tool life and surface roughness modeling in hard turning. The inputs for the ANFIS model were feed and cutting speed at five and four levels, respectively, and tool wear and surface roughness were the outputs. It was found that the ANFIS model slightly outperformed the ANN model in both tool wear and surface roughness prediction. Anil Raj et al. (2016) employed an ANFIS-based model for surface roughness during hard turning with minimal cutting fluid application. Using feed rate, cutting speed and depth of cut as input parameters and surface roughness as output parameter for their model, they managed to create a predictive tool with high level of accuracy.

6.3.3 Response surface methodology

In contrast with soft computing methods, RSM is more similar to regression techniques, but is integrated also in a framework that includes design of experiments method for effectively reducing the number of experiments needed and an optimization method for the determination of optimum process parameters. A selection of relevant papers will be discussed hereafter.

Noordin et al. (2004) conducted experiments concerning the performance of coated carbide tool, during the hard turning of AISI 1045 steel. Cutting speed, depth of cut, and side cutting edge angle were selected as inputs whereas surface roughness and tangential component of force were the outputs. The experiments were designed using a Central Composite Design (CCD)—16 experiments in total—and values of R^2, R^2-adjusted, and R^2-predicted exceeded 85% for all RSM models. Lalwani et al. (2008) investigated the effect of cutting parameters on cutting forces and surface roughness in hard turning of MDN250 steel using RSM models. Horng et al. (2008) investigated the machinability of Hadfield steel during hard turning with ceramic cutting tool. They conducted 30 experiments using a CCD design and apart from typical process parameters they included tool corner radius as input parameter, whereas maximum flank wear (VBmax) and Surface roughness Ra were the output parameters. The RSM models for VBmax and Ra had R^2 and R^2-adjusted values over 90%, and despite R^2-predicted being near 70%, the models were considered adequate as the lack of fit was not significant.

Gaitonde et al. (2009) carried out experimental investigations for the hard turning process of AISI 420 steel. Apart from cutting speed and feed rate, they included machining time as input parameters and conducted experiments based on L27 Taguchi array. The outputs of the model were machining force and power, specific cutting force, Ra, and tool wear (VB). The developed RSM models were found to be statistically significant for all outputs and all models exhibited high values of R^2.

Bouacha et al. (2010) conducted hard turning experiments on AISI 52100 workpieces with CBN tools. In total, 27 machining experiments designed by L_{27} Taguchi array were conducted and surface roughness parameters Ra, Rt, Rz as well as force components were recorded for each experiment. These quantities were employed as input and output parameters for the RSM models, respectively. All RSM models exhibited high values of R^2 and R^2-adjusted (>90%) and normality tests indicated the appropriateness of these models for the prediction of surface roughness and force during hard turning. Paiva et al. (2012) investigated the optimization of hard turning process of AISI 52100 steel with ceramic tools using both actual variables, i.e., feed, depth of cut, and cutting speed, as well as noise parameters, i.e., hardness and tool wear, for the RSM model. They conducted nineteen experiments, using a CCD design and recorded mean and variance of five surface roughness quantities, namely, Ra, Rz, Ry, Rt, and Rq. The RSM model was integrated into a general optimization methodology and exhibited R^2-adjusted values over 80%. Rao et al. (2012) carried out experimental studies of optimization of hard milling process of EN31 steel concerning material removal rate and cutting temperatures and developed RSM models from these experiments.

Hessainia et al. (2013) created a model for surface roughness prediction during hard turning based on process parameters and cutting tool vibrations. More specifically, 27 experiments with various combinations of cutting speed, feed, and depth of cut, designed using an L_{27} Taguchi orthogonal array were carried out and roughness and vibration values were recorded. As for Ra prediction the R^2 value was 99.9% and 96.4% for Rt whereas adjusted R^2 value was 99.4% and 84.6%, respectively, thus indicating the goodness of fit of the models. Using RSM models and desirability function, the optimum set of parameters for minimization of Ra and Rt was determined. Sahoo and Sahoo (2013) conducted hard machining experiments, concerning inserts with hard surface coatings. The RSM model which was created had three inputs, namely, depth of cut, feed, and cutting speed and outputs, flank wear, Ra and Rz; a total of 64 sets of data were used. From the results of the RSM model, the main effects and significant parameters were determined and R^2-adjusted values over 80% for each model indicated the level of accuracy of the models, which was superior to that of simpler regression models.

Das et al. (2014) investigated the effect of process parameters on process performance characteristics during hard machining of AISI 4340 steel. They conducted 9 experiments based on Taguchi L_9 array with three parameters, i.e., cutting speed, feed, depth of cut, at three levels and observed the results on surface roughness, machining force, and tool wear. The RSM models were tested for normality using the Anderson-Darling test and afterward they were used for the optimization of the process. Bouzid et al. (2015) created RSM models for the prediction of surface roughness and the three components of machining force during hard turning. Their models were checked for normality of error residuals and their R^2 values exceeded 90% in all cases.

Mia and Dhar (2016b) studied the cutting temperatures during hard turning. In their RSM model, they considered cutting speed, feed rate, as well as hardness for inputs and the outputs of the model were temperatures during dry hard turning and HPC hard turning. They conducted 54 experiments, i.e., full factorial design with 27 experiments for dry and HPC cutting, and their models were found to exhibit very high level of accuracy ($R^2 > 98\%$ for all cases). Ngygen and Hsu (2016) conducted hard milling experiments with JIS SKD61 steel workpieces and developed models for the prediction of surface roughness. The RSM model included cutting speed, spindle speed, feed, axial depth of cut, and material hardness as input, whereas Ra was the only output. Their model was able to achieve a R^2 value greater than 90% and thus it was subsequently used for the optimization of the process and the identification of significant factors. Bensouilah et al. (2016) conducted a study concerning the performance of cutting tools during hard turning process of AISI D3 steel. They planned their experiments using Taguchi method and analyzed the results using ANOVA, signal-to-noise ratio, and RSM. Input parameters of the model were cutting speed, feed rate, and depth of cut while surface roughness and cutting force components were used as outputs. Separate models were created for the coated and the uncoated tools. Interactions between parameters, significant factor, and optimum set of parameters were determined.

6.4 Conclusions

In this chapter, a concise literature study of modeling methods for hard machining processes has been conducted with a view to present the already established methods on this subject and provide the interested readers who wish to conduct relevant studies with a strong and sufficient background. More specifically:

- The most widely employed machining processes such as hard turning, hard milling, and hard drilling, as well as other less frequently used ones were briefly presented and their particularities were discussed. Furthermore, phenomena occurring during these processes such as residual stresses or microstructure alterations were also described.
- The use of numerical simulation models for the investigation of hard machining processes was described and several works on model with various levels of complexity were discussed, namely, thermal models and thermomechanical models. In the case of thermomechanical models, an emphasis was also put on the presentation of specialized flow stress models and microstructure-related phenomena modeling, which is considered to be the state of the art in this field of modeling.
- Finally, several works, pertinent to the promising and particularly helpful field of soft computing and statistical methods were presented with a view to inform the reader of the applications of these methods regarding hard machining processes, as well as demonstrate the possibilities of these methods, which can be useful to be employed by the readers in future works.

References

Abdullah, A.A., Naeem, U.J., Xiong, C., 2012. Estimation and optimization cutting conditions of surface roughness in hard turning using Taguchi approach and artificial neural network. Adv. Mater. Res. 463–464, 662–668.

Agrawal, A., Goel, S., Rashid, W.B., Price, M., 2015. Prediction of surface roughness during hard turning of AISI 4340 steel (69HRC). Appl. Soft Comput. J. 30, 279–286.

Akkus, H., Asilturk, I., 2011. Predicting surface roughness of AISI 4140 Steel in hard turning process through artificial neural network, fuzzy logic and regression models. Sci. Res. Essays 6 (13), 2729–2736.

Al Hazza, M., Adesta, E.Y.T., 2011. Flank wear modeling in high speed hard turning by using artificial neural network and regression analysis. Adv. Mater. Res. 264–265, 1097–1101.

Anil Raj, R., Anand, M.D., Wins, K.L.D., Varadarajan, A.S., 2016. ANFIS based model for surface roughness prediction for hard turning with minimal cutting fluid application. Indian. J. Sci. Technol. 9(13).

Arrazola, P.J., Ozel, T., 2008. Numerical modelling of 3D hard turning using arbitrary Lagrangian Eulerian finite element method. Int. J. Mach. Mach. Mater. 3 (3), 238–249.

Arulraj, J.G.A., Wins, K.L.D., Raj, A., 2014. Artificial neural network assisted sensor fusion model for predicting surface roughness during hard turning of H13 steel with minimal cutting fluid application. Prog. Mater Sci. 5, 2338–2346.

Astakhov, V.P., 2011. Machining of hard materials—definitions and industrial applications. In: Davim, J.P. (Ed.), Machining of Hard Materials. Springer-Verlag, London, pp. 1–32.

Bagci, E., 2011. 3-D numerical analysis of orthogonal cutting process via mesh-free method. Int. J. Phys. Sci. 6 (6), 1267–1282.

Bapat, P.S., Dhikale, P.D., Shinde, S.M., Kulkarni, A.P., Chinchanikar, S.S., 2015. A numerical model to obtain temperature distribution during hard turning of AISI 52100 steel. Mater. Today: Proc. 2, 1907–1914.

Bartarya, G., Choudhury, S.K., 2012. State of the art in hard turning. Int. J. Mach. Tools Manuf. 53 (1), 1–14.

Beatrice, B.A., Kirubakaran, E., Thangaiah, P.R.J., Wins, K.L.D., 2014. Surface roughness prediction using artificial neural network in hard turning of AISI H13 steel with minimal cutting fluid application. Proc. Eng. 97, 205–211.

Bensouilah, H., Aouici, H., Meddour, I., Yallese, M.A., Mabrouki, T., Girandin, F., 2016. Performance of coated and uncoated mixed ceramic tools in hard turning process. Measurement 82, 1–18.

Bhemuni, V.P., Chalamalasetti, S.R., 2013. A review on hard turning by using design of experiments. Int. J. Manuf. Sci. Prod. 13 (3), 209–219.

Bouacha, K., Yallese, M.A., Mabrouki, T., Rigal, J.F., 2010. Statistical analysis of surface roughness and cutting forces using response surface methodology in hard turning of AISI 52100 bearing steel with CBN tool. Int. J. Refract. Met. Hard Mater 28, 349–361.

Bouzid, L., Yallese, M.A., Chaoui, K., Mabrouki, T., Boulanouar, L., 2015. Mathematical modeling for turning on AISI 420 stainless steel using surface response methodology. Proc. Inst. Mech. Eng. B J. Eng. Manuf. 229 (1), 45–61.

Cakir, M.C., Isik, Y., 2005. Finite element analysis of cutting tools prior to fracture in hard turning operations. Mater. Des. 26, 105–112.

Campos, P.H.S., Ferreira, J.R., de Paiva, A.P., Balestrassi, P.P., Davim, J.P., 2013. Modeling and optimization techniques in machining of hardened steels: a brief review. Rev. Adv. Mater. Sci. 34, 141–147.

Campos, P., Davim, J.P., Ferreira, J.R., Paiva, A.P., Balestrassi, P.P., 2014. The machinability of hard materials—a review. In: Davim, J.P. (Ed.), Machinability of Advanced Materials. John Wiley & Sons, Inc, pp. 145–173.

Caruso, S., Di Renzo, S., Umbrello, D., Jayal, A.D., Dillon, O.W., Jawahir, I.S., 2011. Finite element modeling of microstructural changes in hard turning. Adv. Mater. Res. 223, 960–968.

Chinchanikar, S., Choudhury, S.K., 2015. Machining of hardened steel—experimental investigations, performance modeling and cooling techniques: a review. Int. J. Mach. Tools Manuf. 89, 95–109.

Cica, D., Sredanovic, B., Kramar, D., 2015. Modeling of tool life and surface roughness in hard turning using soft computing techniques: a comparative study. Int. J. Mater. Prod. Technol. 50 (1), 49–64.

Das, S.D., Nayak, R.P., Dhupal, D., Kumar, A., 2014. Surface roughness, machining force and flank wear in turning of hardened AISI 4340 steel with coated carbide insert: cutting parameters effects. Int. J. Automot. Eng. 4 (3), 758–767.

Ding, H., Shin, Y.C., 2013. Multi-physics modeling and simulations of surface microstructure alteration in hard turning. J. Mater. Process. Technol. 213, 877–886.

Dogra, M., Sharma, V.S., Sachdeva, A., Suri, N.M., Dureja, J.S., 2010. Tool wear, Chip Formation and Workpiece Surface issues in CBN hard turning: A review. Int. J. Precis. Eng. Manuf. 11 (2), 341–358.

El-Magd, E., Treppmann, C., 2000. Mechanical behavior of AA7075, Ck45N and TiAl6V4 at high strain rates. Materialsweek 1–8. Munich.

Fergani, O., Pan, Z., Liang, S.Y., Atmani, Z., Welo, T., 2016. Microstructure texture prediction in machining processes. Procedia CIRP 46, 595–598.

Gaitonde, V.N., Karnik, S.R., Figueira, L., Davim, J.P., 2009. Analysis of machinability during hard turning of cold work tool steel (Type: AISI D2). Mater. Manuf. Process. 24, 1373–1382.

Goldthorpe, B.D., Church, P., 1997. The effect of algorithm form on deformation and instability in tension. Proceedings of the EURODYMAT 97, J. de Physique IV, Colloque C8 7, 753–759.

Grzesik, W., 2008. Machining of hard materials. In: Davim, J.P. (Ed.), Machining: Fundamentals and Recent Advances. Springer Verlag, London, pp. 97–126.

Grzesik, W., 2011. Mechanics of cutting and chip formation. In: Davim, J.P. (Ed.), Machining of Hard Materials. Springer Verlag, London, pp. 87–114.

Guo, Y.B., Liu, C.R., 2002. 3D FEA modeling of hard turning. J. Manuf. Sci. Eng. 124, 189–199.

Guo, Y.B., Yen, D.W., 2004. A FEM study on mechanisms of discontinuous chip formation in hard machining. J. Mater. Process. Technol. 155–156, 1350–1356.

Hashmi, K., El Baradie, M.A., Ryan, M., 1998. Fuzzy logic based intelligent selection of machining parameters. Comput. Ind. Eng. 35 (3-4), 571–574.

Hessainia, Z., Belbah, A., Yallese, M.A., Mabrouki, T., Rigal, J.F., 2013. On the prediction of surface roughness in the hard turning based on cutting parameters and tool vibrations. Measurement 46 (5), 1671–1681.

Horng, J.-T., Liu, N.-M., Chiang, K.-T., 2008. Investigating the machinability evaluation of Hadfield steel in the hard turning with Al_2O_3/TiC mixed ceramic tool based on the response surface methodology. J. Mater. Process. Technol. 208, 532–541.

Hua, J., Umbrello, D., Shivpuri, R., 2006. Investigation of cutting conditions and cutting edge preparations for enhanced compressive subsurface residual stress in the hard turning of bearing steel. J. Mater. Process. Technol. 171, 180–187.

Huang, Y., Liang, S.Y., 2003. Cutting forces modeling considering the effect of tool thermal property-application to CBN hard turning. Int. J. Mach. Tools Manuf. 43, 307–315.

Jivishov, V., Rzayev, E., 2016. Influence of material models used in finite element modeling on cutting forces in machining. Mater. Sci. Eng. 142, 012072. IOP Conference Series.

Kim, D.M., Bajpai, V.B., Kim, B.H., Park, H.W., 2015. Finite element modeling of hard turning process via a micro-textured tool. Int. J. Adv. Manuf. Technol. 78, 1393–1405.

Klocke, F., Doebbeler, B., Buchkremer, S., 2016. On the applicability of the concept of process signatures to hard turning. Procedia CIRP 45, 7–10.

Koppka, F., Sahlan, H., Sartkulvanich, P., Altan, T., 2001. Experimental determination of flow stress data for FEM simulation of machining operations, ERC Report, ERC/NSM-01-R-63.

Kountanya, R., Al-Zkeri, I., Altan, T., 2009. Effect of tool edge geometry and cutting conditions on experimental and simulated chip morphology in orthogonal hard turning of 100Cr6 steel. J. Mater. Process. Technol. 209, 5068–5076.

Kryzhanivskyy, V., Bushlya, V., Gutnichenko, O., Petrusha, I.A., Staehl, J.E., 2015. Modeling and experimental investigation of cutting temperature when rough turning hardened tool steel with PCBN tools. Procedia CIRP 31, 489–495.

Kundrák, J., Szabó, G., Markopoulos, A.P., 2016. Numerical investigation of the influence of tool rake angle on residual stresses in precision hard turning. Key Eng. Mater. 686, 68–73.

Kundrák, J., Gyáni, K., Tolvaj, B., Pálmai, Z., Tóth, R., Markopoulos, A.P., 2017. Thermotechnical modelling of hard turning: a computational fluid dynamics approach. Simul. Modell. Pract. Theory 70, 52–64.

Lalwani, D.I., Mehta, N.K., Jain, P.K., 2008. Experimental investigations of cutting parameters influence on cutting forces and surface roughness in finish hard turning of MDN250 steel. J. Mater. Process. Technol. 206, 167–179.

Mamalis, A.G., Kundrák, J., Markopoulos, A., Manolakos, D.E., 2008. On the finite element modelling of high speed hard turning. Int. J. Adv. Manuf. Technol. 38 (5-6), 441–446.

Mankova, I., Kovac, P., Kundrák, J., Beno, J., 2011a. Finite element analysis of hardened steel cutting. J. Prod. Eng. 14 (1), 7–10.

Mankova, I., Kovac, P., Kundrák, J., Beno, J., 2011b. Outline of FEM simulation and modeling of hard turning process. Acta Mechanica Slovaca 15 (3), 14–21.

Markopoulos, A.P., Habrat, W., Galanis, N.I., Karkalos, N.E., 2016. Modeling and optimization of machining with the use of statistical methods and soft computing. In: Davim, J.P. (Ed.), Design of Experiments in Production Engineering. Springer International Publishing, pp. 39–88.

Mia, M., Dhar, N.R., 2016a. Prediction of surface roughness in hard turning under high pressure coolant using Artificial neural network. Measurement 92, 464–474.

Mia, M., Dhar, N.R., 2016b. Response surface and neural network based predictive models of cutting temperature in hard turning. J. Adv. Res. 7 (6), 1035–1044.

Nakayama, K., Arai, M., Kanda, T., 1988. Machining characteristics of hard materials. CIRP Ann. Manuf. Technol. 37 (1), 89–92.

Ng, E.G., Aspinwall, D.K., 2002. Modelling of hard part machining. J. Mater. Process. Technol. 127, 222–229.

Ngygen, H.T., Hsu, Q.C., 2016. Surface roughness analysis in the Hard Milling of JIS SKD 61 alloy steel. Appl. Sci. 6, 172.

Noordin, M.Y., Venkatesh, V.C., Sharif, S., Elting, S., Abdullah, A., 2004. Application of response surface methodology in describing the performance of coated carbide tools when turning AISI 1045 steel. J. Mater. Process. Technol. 145, 46–58.

Osorio, J.M.A., Velasco, O.G.D., Rodriguez, C.J.C., 2007. Cutting tool geometry suggestions based on a fuzzy logic model. In: Proceedings of COBEM 2007, 19th International Congress of Mechanical Engineering.

Oxley, P.L.B., 1989. The Mechanics of Machining: An Analytical Approach to Assessing Machinability. E. Horwood, Chichester.

Özel, T., Karpat, Y., 2005. Predictive modeling of surface roughness and tool wear in hard turning using regression and neural networks. Int. J. Mach. Tools Manuf 45, 467–479.

Paiva, A.P., Campos, P.H., Ferreira, J.R., Lopes, L.G.D., Paiva, E.J., Balestrassi, P.P., 2012. A multivariate robust parameter design approach for optimization of AISI 52100 hardened steel turning with wiper mixed ceramic tool. Int. J. Refract. Met. Hard Mater. 30 (1), 152–163.

Poulachon, G., Moisan, A., Jawahir, I.S., 2001. On modeling the influence of thermo-mechanical behavior in chip formation during hard turning of 100Cr6 bearing steel. CIRP Ann. Manuf. Technol. 50 (1), 31–36.

Quiza, R., Figueira, L., Davim, J.P., 2008. Comparing statistical models and artificial neural networks on predicting the tool wear in hard machining D2 AISI steel. Int. J. Adv. Manuf. Technol. 37, 641–648.

Rana, K., Rinaldi, S., Imbrogno, S., Rotella, G., Umbrello, D., M'Saoubi, R., Ayvar-Soberanis, S., 2016. 2D FE prediction of surface alteration of Inconel 718 under machining condition. Procedia CIRP 45, 227–230.

Rao, L.P., Kumar, T.J.A., Rao, T.B., 2012. Modelling and multi-response optimization of hard milling process based on RSM and GRA approach. Int. J. Eng. Res.Technol. 1 (9), 1–6.

Sahoo, A.K., Sahoo, B., 2013. Performance studies of multilayer hard surface coatings (TiN/TiCN/Al_2O_3/TiN) of indexable carbide inserts in hard machining: Part II (RSM, grey relational and techno economical approach). Measurement 46, 2868–2884.

Saini, S., 2015. Regression vs neuro-fuzzy model: A comparison for residual stress, surface roughness, tool wear during hard turning. In: Proceedings of IRF International Conference, Mumbai, India, pp. 17–22.

Senthilkumar, N., Tamizharasan, T., 2015. Flank wear and surface roughness prediction in hard turning via artificial neural network and multiple regressions. Aust. J. Mech. Eng. 13 (1), 31–45.

Sharma, V.S., Dhiman, S., Sehgal, R., Sharma, S.K., 2008. Estimation of cutting forces and surface roughness for hard turning using neural networks. J. Intell. Manuf. 19, 473–483.

Shatla, M., Kerk, C., Altan, T., 2001. Process modeling in machining. Part I. Determination of flow stress data. Int. J. Mach. Tools Manuf. 41, 1511–1534.

Shibab, S.K., Khan, Z.A., Mohammad, A., Siddiquee, A.N., 2014. A review of turning of hard steels used in bearing and automotive applications. Prod. Manu. Res. 2 (1), 24–49.

Steinberg, D.J., Cochran, S.G., Guinan, M.W., 1980. A constitutive model for metals applicable at high-strain rate. J. Appl. Phys. 51 (3), 1498–1504.

Sukaylo, V.A., Kaldos, A., Krykovsky, G., Lierath, F., Emmer, T., Pieper, H.J., Kundrák, J., Bana, V., 2004. Development and verification of a computer model for thermal distortions in hard turning. J. Mater. Process. Technol. 155–156, 1821–1827.

Szabó, G., Kundrák, J., 2012. Numerical research of the plastic strain in hard turning in case of orthogonal cutting. Key Eng. Mater. 496, 162–167.

Takacs, M., Farkas, B.Z., 2014. Hard cutting of AISI D2 steel. In: Proceedings of the 3rd International Conference on Mechanical Engineering and Mechatronics, Prague, Czech Republic. paper no. 176.

Tamilarasan, A., Rajamani, D., Renugambal, A., 2015. An approach on fuzzy and regression modeling for hard milling process. Appl. Mech. Mater. 813–814, 498–504.

Toenshoff, H.K., Denkena, B., 2013. Basics of Cutting and Abrasive Processes, Lecture Notes in Production Engineering. Springer-Verlag, Berlin; Heidelberg.

Uhlmann, E., Mahnken, R., Ivanov, I.M., Cheng, C., 2015. A novel finite element approach to modeling hard turning in due consideration of the viscoplastic asymmetry effect. Procedia CIRP 31, 471–476.

Umbrello, D., Hua, J., Shivpuri, R., 2004. Hardness-based flow stress and fracture models for numerical simulation of hard machining AISI 52100 bearing steel. Mater. Sci. Eng. A 374, 90–100.

Umbrello, D., Rizzuti, S., Outeiro, J.C., Shivpuri, R., M'Saoubi, R., 2008. Hardness-based flow stress for numerical simulation of hard machining AISI H13 tool steel. J. Mater. Process. Technol. 199, 64–73.

Umbrello, D., Jayal, A.D., Caruso, S., Dillon, O.W., Jawahir, I.S., 2010. Modeling of white and dark layer formation in hard machining of AISI 52100 bearing steel. Mach. Sci. Technol. 14 (1), 128–147.

Wang, X., Yuang, Y., 2009. Optimized recurrent neural network-based tool wear modeling in hard turning. Transactions of NAMRI/SME. 27.

Wong, S.V., Hamouda, A.M.S., El Baradie, M.A., 1999. Generalized fuzzy model for metal cutting data selection. J. Mater. Process. Technol. 89–90, 310–317.

Yan, H., Hua, J., Shivpuri, R., 2005. Numerical simulation of finish hard turning for AISI H13 die steel. Sci. Technol. Adv. Mater. 6, 540–547.

Zerilli, F.J., Armstrong, R.W., 1987. Dislocation mechanics-based constitutive relations for material dynamics calculations. J. Appl. Phys. 61 (5), 1816–1825.

Zuperl, U., Cus, F., 2015. End-milling force control system with surface roughness monitoring. Mach. Des. 7 (3), 85–88.

Multiresponse optimization in wire electric discharge machining (WEDM) of HCHCr steel by integrating response surface methodology (RSM) with differential evolution (DE)

7

V.N. Gaitonde, M. Manjaiah†, S. Maradi*, S.R. Karnik*, P.M. Petkar*, J. Paulo Davim‡*
*B.V.B. College of Engineering and Technology, Hubli, India, †GeM Lab, Nantes, France, ‡University of Aveiro, Aveiro, Portugal

7.1 Introduction

Since 19th century, there has been number of improvements in the manufacturing industries. Major revolutions appeared at the computer numerical control (CNC) for efficient and low cost productions. Industries always establish different manufacturing processes to increase the flexibility and adaptability for productivity improvement. High carbon high chromium (HCHCr) steels are the major materials used in tool and die making industry. There is an incredible demand for HCHCr steel in cold forming molds, press tools, and die making industries mainly due to excellent wear resistance, high compressive strength, greater dimensional stability, and high hardness (Novotny, 2001). High carbon and chromium contents of this alloy increase the mechanical strength as well as hardness of the material due to the formation of ultrahard and abrasive carbides, resulting into high cutting temperature, larger stresses causing greater tool wear, as well as cutting tool failure, particularly during the conventional machining (Bhattacharya et al., 2015). Owing to poor machinability and surface integrity, HCHCr is considered as difficult to cut material (Jomaa et al., 2011).

The chain links are manufactured with high carbon steel material by forming process. These technologies rely on the application of high thermal and mechanical loading on the forming tools and the microstructural properties. The manufacturers therefore unavoidably needed to adopt enhanced technologies to achieve better surface quality, dimensional accuracy, higher production rate, and lower cost of production (Coldwell et al., 2003). Hence, in the current study, an attempt has been made to machine the punch profile for chain link component using wire electro discharge machining (WEDM) process. It is a process for machining hard-to-cut materials and complex surfaces. The wire electro discharge machining (WEDM)

Computational Methods and Production Engineering. http://dx.doi.org/10.1016/B978-0-85709-481-0.00007-0

is a thermoelectric process to machine electrically conductive hard-to-cut material and gained popularity in machining complex shapes with desired accuracy and dimensions.

Several researchers attempted to improve the performance characteristics such as material removal rate (MRR), surface roughness, surface and subsurface properties in WEDM. But full potential utilization of this process is still not completely solved due to its complexity, stochastic nature, and in addition, more number of process variables is involved in this process. Lodhi and Agarwal (2014) performed L_9 orthogonal array of experiments to optimize the process parameters during WEDM of AISI D3 steel. The discharge current and pulse on time are found to be the most significant factors affecting the surface roughness. Mir et al. (2012) performed modeling and analysis on powder mixed EDM of H11 steel to improve the machining surface quality. The effects of process parameters such as pulse on time, discharge current, and concentration of aluminum powder in dielectric fluid were studied on the performance characteristics. The improvement in surface quality was observed with the addition of Al powder into the dielectric fluid. Purohit et al. (2015) optimized the electric discharge machining parameters of M2 tool steel using Grey relational approach. The material removal rate; electrode wear ratio; and overcut with respect to tool rotation speed, voltage, and spark time were optimized using L_9 orthogonal array of experiments. The electrode rotation speed was found to be the most significant factor followed by voltage and spark time. Ramakrishnan and Karunamoorthy (2006) developed artificial neural network (ANN) modeling to predict the surface roughness and MRR during WEDM of Inconel 718. Further, the surface roughness and MRR were concurrently optimized using multiresponse signal-to-noise ratio (MRSN). Singh and Pradhan (2014) made an attempt to investigate the effect of WEDM process parameters on MRR, surface roughness, and cutting rate. AISI D2 steel specimens were machined with brass wire electrode using L_{27} orthogonal array of experiments. The pulse on time and pulse off time were the major factors affecting the cutting rate and MRR, while the surface roughness was affected by pulse on time and servo voltage.

The prior works on WEDM indicate that most of the studies concentrated on the influence of process parameters on surface roughness and MRR during machining of different steels. The HCHCr die steel is the best suitable material for producing punches, with high compressive strength, high impact strength, high wear resistance, better hardening properties, and good resistance to tempering back. In industries, especially, for shearing punches, the profiles are cut by WEDM process after hardening to 62–65 HRc. For such punches, it is very difficult to machine the profile with traditional machining. Despite extensive research on WEDM process parameters, determining the desirable operating conditions during WEDM of HCHCr steel in industrial settings still relies on operator skills and trial-and-error methods.

Mathematical model developed through the response surface methodology (RSM) using design of experiments (DOE) is found to be proficient procedure (Montgomery, 2003; Myers et al., 2009). The matrix experiments using DOE/orthogonal array (OA) enable to study the outcome of numerous parameters concurrently with least number of experiments with efficient time utilization and reduced material consumption.

Depending on the number of process parameters and levels, a suitable DOE or OA is selected. Each column of the array represents the process parameter and its desired level in experimental conditions. Conversely, the genetic algorithms (Gaitonde et al., 2008a,b, 2012a; Karnik et al., 2007), particle swarm optimization (Gaitonde et al., 2012b,c; Gaitonde and Karnik, 2012; Mata et al., 2013) simulated annealing (Karnik et al., 2013), gravitational search algorithm (Esmat et al., 2009), and differential evolution (Price et al., 2005) are well-known multiresponse optimization tools applied in various engineering fields including the machining operations. Hence, the aim of the present study is develop the mathematical models using RSM for the WEDM characteristics and then to simultaneously optimize the proposed WEDM characteristics of HCHCr steel using (DE) algorithm.

7.2 Experimental work

The experiments were performed in *Electronic* WEDM (*Model: Ecocut*). The composition of HCHCr steel work material used for the experimentation is given in Table 7.1. Before machining, the material was subjected to hardening (up to 62HRc) to attain the required strength and hardness. After hardening, the sides were ground to get two pairs of right-angled surfaces, which is necessary for appropriate mounting of the workpiece on WEDM work bed. The size of the workpiece is 125 mm × 50 mm × 15 mm. The brass wire with 0.25 mm diameter was used for machining the HCHCr steel.

In the present work, three parameters, namely, pulse on time (T_{on}), pulse off time (T_{off}), and wire feed (WF) are considered and the ranges of the parameters were selected based on the earlier studies of the authors (Narendranath et al., 2013; Manjaiah et al., 2014, 2015, 2016a,b). The selected process parameters for the experimentation and their levels are given in Table 7.2 and the peak current of 12A, servo voltage of 20 V, and servo feed of 2250 mm/min were kept constant throughout

Table 7.1 Chemical compositions of HCHCr die steel

Element	C	Cr	Si	Mn	P	S	Ni	Mo	V	W	Fe
Wt (%)	1.93	12.02	0.172	0.342	0.015	0.020	0.113	0.074	0.053	0.007	Bal

Table 7.2 WEDM process parameters and their levels

Code	Parameter	Level		
		I	II	III
A	Pulse on time (μs)	110	115	120
B	Pulse off time (μs)	20	40	60
C	Wire feed (m/min)	4	6	8

Table 7.3 L_{27} orthogonal array and experimental results

Sl. no.	T_{on} (μs)	T_{off} (μs)	WF (m/min)	Ra (μm)	MRR (mm^3/min)	TWR (g/min)
1	110	20	4	1.342	4.173	0.012
2	110	20	6	1.395	3.624	0.063
3	110	20	8	1.440	4.775	0.016
4	110	40	4	1.277	4.043	0.043
5	110	40	6	1.260	4.060	0.084
6	110	40	8	1.302	3.871	0.072
7	110	60	4	1.502	1.936	0.046
8	110	60	6	1.077	1.939	0.083
9	110	60	8	1.587	1.931	0.048
10	115	20	4	2.825	8.912	0.133
11	115	20	6	2.765	10.074	0.192
12	115	20	8	3.027	10.005	0.074
13	115	40	4	2.035	6.000	0.111
14	115	40	6	1.987	6.221	0.045
15	115	40	8	1.957	6.221	0.022
16	115	60	4	1.360	2.772	0.07
17	115	60	6	1.765	2.859	0.061
18	115	60	8	1.727	2.909	0.115
19	120	20	4	3.77	11.142	0.35
20	120	20	6	2.277	11.597	0.3
21	120	20	8	3.195	11.682	0.21
22	120	40	4	3.195	7.093	0.233
23	120	40	6	2.592	7.07	0.225
24	120	40	8	3.03	7.7	0.166
25	120	60	4	2.435	3.568	0.132
26	120	60	6	2.677	3.566	0.015
27	120	60	8	2.807	3.582	0.043

experimentation. Deionized water (2.8 kg/cm^2) was used as the die electric fluid. Taguchi based L_{27} OA of experiments were used for studying the several process parameters in the design and results are presented in Table 7.3. The photographic views of the machining zone and the machined component are shown Fig. 7.1A and B.

The centerline surface roughness (Ra) was measured using *Mitutoyo* surface roughness tester. The surface roughness for each workpiece was measured at four different locations and the average was considered for the analysis. Machining time was determined using stopwatch and width of the cut was measured using *Faro Gauge* (Coordinate Measuring Machine) and shown in Fig. 7.2. The material removal rate (MRR) is calculated by the following equation:

$$\mathrm{MRR} = V_c \times b \times h \left(\frac{\mathrm{mm}^3}{\mathrm{min}}\right) \tag{7.1}$$

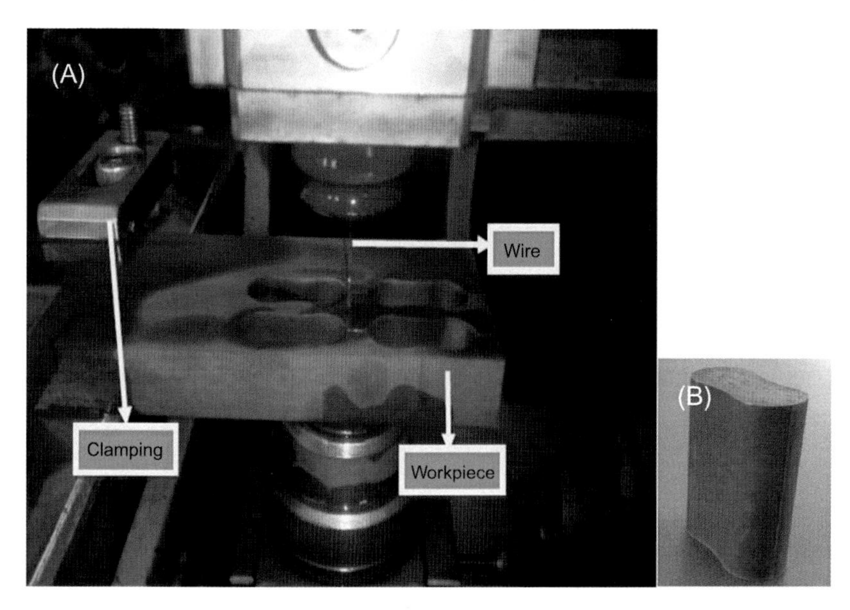

Fig. 7.1 (A) Machining zone and (B) machined component.
Courtesy: B.V.B. College of Engineering and Technology, Hubli, India.

Fig. 7.2 Faro gauge.
Courtesy: B.V.B. College of Engineering and Technology, Hubli, India.

where h is the height (thickness) of job (mm) and V_c is the cutting speed

$$\left(\frac{\text{mm}}{\text{min}}\right)=\frac{\text{Total profile (cutting) length}}{\text{Machining time}} \tag{7.2}$$

b is the width of cut or kerf.

$$(\text{mm})=\frac{\text{Width of pocket}-\text{width of punch}}{2} \tag{7.3}$$

The tool wear rate (TWR) was measured by weight loss method, i.e., initial and final weight of the wire before and after machining.

7.3 Response surface methodology

Response surface methodology (RSM) is a compilation of mathematical and statistical methods, helpful for fitting the models and analyzing the problems in which quite a lot of independent parameters control the dependent parameter (s) (Montgomery, 2003; Myers et al., 2009). The empirical mathematical modeling for any performance characteristic is fitted with the correlating parameters. In RSM, the response surface can be viewed as a surface expressed as (Montgomery, 2003; Myers et al., 2009):

$$Y=\phi(x_1,x_2,x_3,\ldots\ldots,x_i,\ldots\ldots,x_k) \tag{7.4}$$

where Y is the performance characteristic of the system, ϕ is the performance function, x_i is the independent parameter i, and k is the number of parameters.

RSM-based technique consists of choice of an accurate DOE or OA to construct the proper mathematical model for the desired performance characteristic(s) with the best fittings of the experimental/simulated data and the graphical demonstration of interaction effects of parameters. The RSM makes use of regression analysis for the construction of mathematical model of desired characteristic(s) of every course of action. The response surface for the second-order mathematical model consists of linear, nonlinear, and two-factor interaction terms of X_i's parameters and is given by (Montgomery, 2003; Myers et al., 2009):

$$Y=d_0+\sum_{i=1}^{k}d_iX_i+\sum_{i=1}^{k}d_{ii}X_i^2+\sum_{i=1}^{k-1}\sum_{j=i+1}^{k}d_{ij}X_iX_j \tag{7.5}$$

where d_i, d_{ii}, and d_{ij} are the regression coefficients and are computed by (Montgomery, 2003; Myers et al., 2009):

$$d=\left(X^TX\right)^{-1}X^TY \tag{7.6}$$

In the current study, the quadratic models for surface roughness (Ra), material removal rate (MRR), and tool wear rate (TWR) based on RSM have been fitted with pulse on time (T_{on}), pulse off time (T_{off}), and wire feed (WF) as the process parameters. The fitted equations developed through multiple regressions (Montgomery, 2003; Myers et al., 2009) for the proposed WEDM characteristics are given as follows:

$$\begin{aligned}\text{Ra} = & -42.1075 + 0.6020{*}A + 0.0789{*}B - 0.2342{*}C \\ & -0.00164{*}A^2 + 0.000237{*}B^2 + 0.0587{*}C^2 \\ & -0.00109{*}A{*}B - 0.0048{*}A{*}C + 0.00229{*}B{*}C\end{aligned} \tag{7.7}$$

$$\begin{aligned}\text{MRR} = & -526.618 + 8.394{*}A + 1.547{*}B - 0.496{*}C \\ & -0.0324{*}A^2 - 0.000488{*}B^2 + 0.0041{*}C^2 \\ & -0.0141{*}A{*}B + 0.006{*}A{*}C - 0.00435{*}B{*}C\end{aligned} \tag{7.8}$$

$$\begin{aligned}\text{EWR} = & \, 8.39614 - 0.19784{*}A + 0.06892{*}B \\ & +0.33435{*}C + 0.00110{*}A^2 - 0.000005{*}B^2 \\ & -0.00333{*}C^2 - 0.00063{*}A{*}B - 0.00276{*}A{*}C \\ & +0.00032{*}B{*}C\end{aligned} \tag{7.9}$$

where A is pulse on time in μs, B is pulse off time in μs, and C is wire feed in m/min, Ra in μm, MRR in mm^3/min, and TWR in g/min.

The adequacy of the developed mathematical models was statistically checked through the analysis of variance (ANOVA) (Montgomery, 2003; Myers et al., 2009). The ANOVA consists of sum of squares, degrees of freedom, mean square, and F-ratio. The sum of squares is generally contributed from the regression model and residual error. Mean square is the ratio of the sum of the squares to the degree of freedom and the F-ratio is the ratio of the mean square of the regression model to the mean square of the residual error. If the F-ratio of the regression model is more than the standard tabulated value (F-Table) for a known confidence interval, then the model is considered as adequate. The ANOVA results for the current investigation at 95% confidence interval are summarized in Table 7.4. As seen from Table 7.4, the fitted mathematical models are significant at 95% confidence interval as F-ratio of all the models is greater than 4.39 [F-Table (9,17,0.05)]. Table 7.4 also gives the R^2 and R^2_{adj} values for the constructed models, which obviously show the competence for the proposed mathematical models of WEDM characteristics of HCHCr steel.

7.4 Results and discussion

The effects of the process parameters were analyzed for the proposed WEDM characteristics (Figs. 7.3, 7.4, 7.7, 7.8, 7.10, and 7.11) using Design-Expert software. The developed Eqs. (7.7–7.9) are used to predict the surface roughness (Ra), material removal rate (MRR), and tool wear rate (TWR) respectively, by substituting the values of pulse on time (T_{on}), pulse off time (T_{off}), and wire feed (WF) within the restrictions of the considered parameters. The influence of the identified parameters on proposed WEDM performance characteristics is demonstrated in Figs. 7.5, 7.9, and 7.12. Here, every planned characteristic is plotted as a

Table 7.4 **Summary of ANOVA for surface roughness, MRR, and TWR models**

	Sum of squares		Degrees of freedom		Mean square				
WEDM characteristic	**Regression**	**Residual**	**Regression**	**Residual**	**Regression**	**Residual**	**F-ratio**	R^2 (%)	R^2_{adj} (%)
Ra	12.6901	2.4074	9	17	1.4100	0.4100	9.96	84.50	75.6
MRR	247.680	5.730	9	17	27.520	0.337	81.65	97.74	96.5
TWR	0.182814	0.026845	9	17	0.020313	0.001579	12.86	87.20	80.4

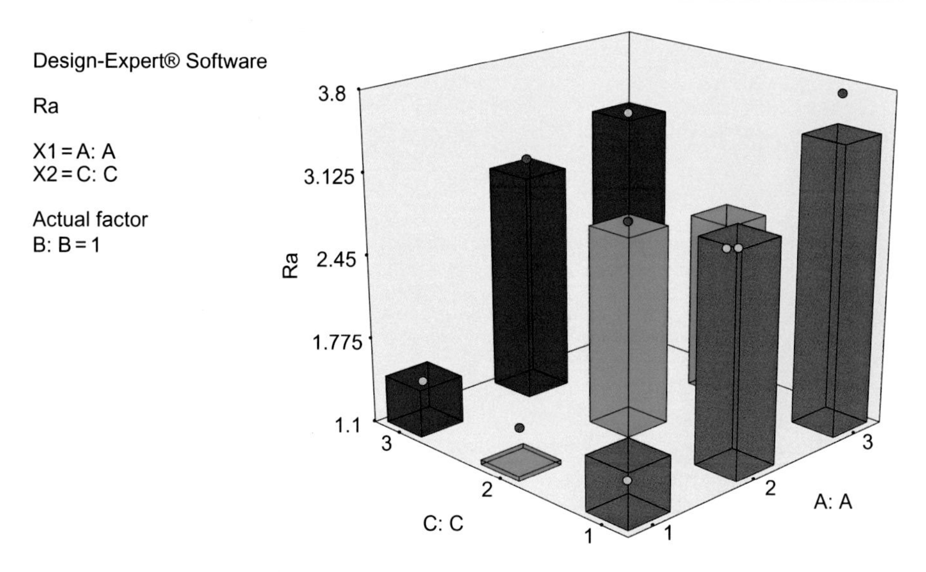

Fig. 7.3 Effect of pulse on time and wire feed against the surface roughness.

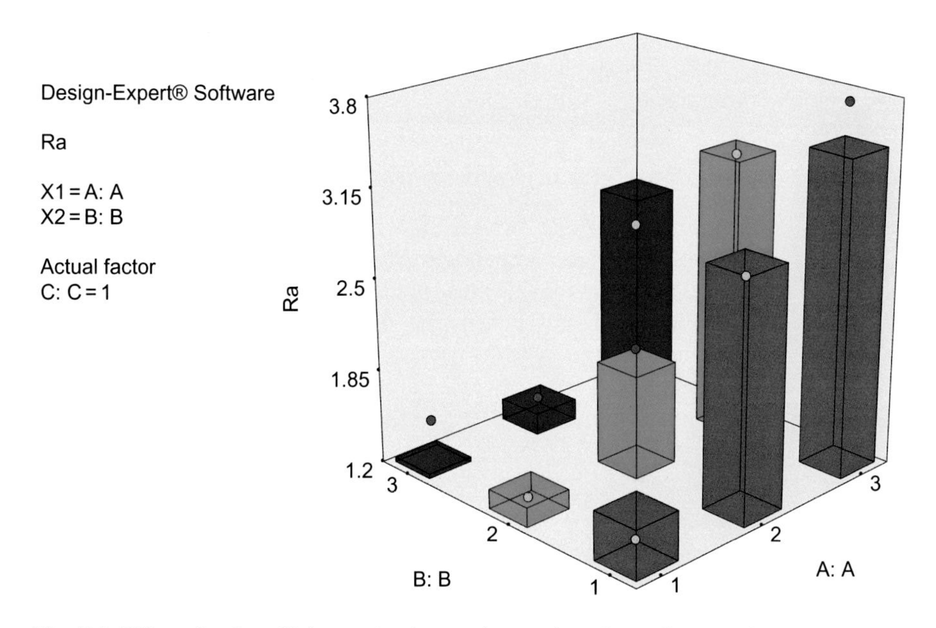

Fig. 7.4 Effect of pulse off time and pulse on time against the surface roughness.

function of pulse off time (T_{off}) with hold values for three different combinations of pulse on time (T_{on}), i.e., 110, 115, and 120 μs and for three different values of wire feed (WF), i.e., at 4, 6, and 8 m/min. It is quite evident from Figs. 7.5, 7.9, and 7.12 that there exists substantial interaction influence between the variables on surface roughness (Ra), material removal rate (MRR), and tool wear rate (TWR).

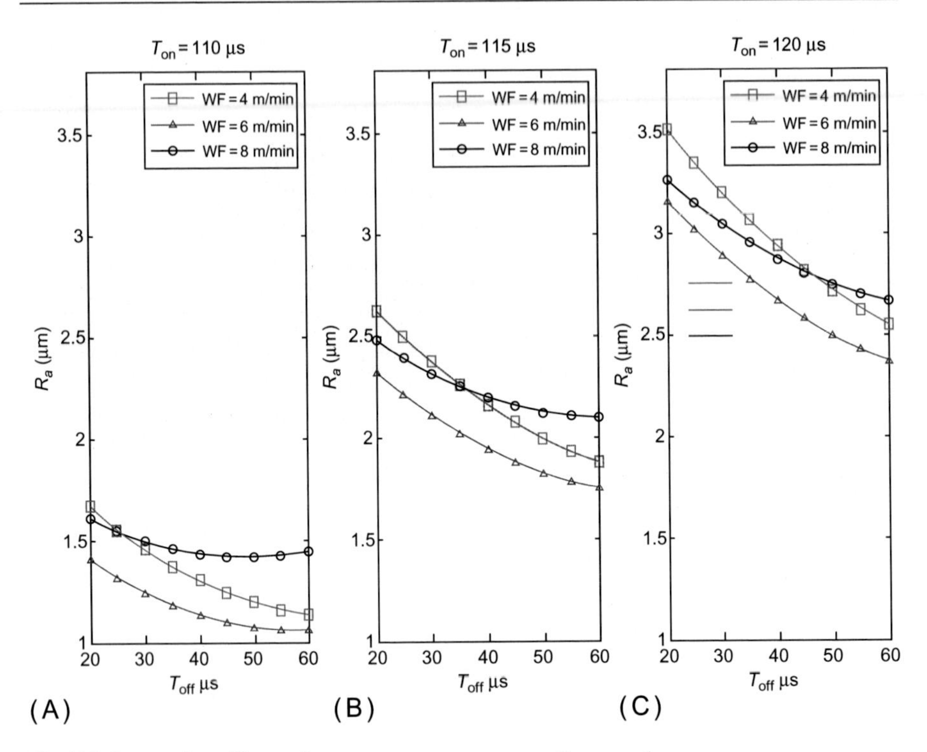

Fig. 7.5 Interaction effects of process parameters on surface roughness.

7.4.1 Analysis of surface roughness (Ra)

Fig. 7.3 shows the influence of pulse on time and wire feed on surface roughness. It is observed from the plot that increase of pulse on time increases the surface roughness. This is due to the fact that the increase in pulse duration causes more heat transfer into the sample and melting isotherms penetrating deeper into the material surface. The melted molten extends further onto the materials surface and this in turn produces a recast layer thickness. As a result, overheating causes for entrapment of gasses leading to the formation of blowholes on the surface. This may further cause for the increase in surface roughness. The energy content in each spark is more violent in the process, thereby produces deeper craters on the surface leading to increased surface roughness. The similar results were also observed by the authors (Hascalik and Caydas, 2007; Lee and Li, 2001). Similarly, the surface roughness is higher at lower wire feed; further increase in wire feed decreases the surface roughness, again it increased at higher wire feed rate. This is because at lower wire feed (4 m/min) continuous sparks are generated and thereby producing lower cutting speed. At lower cutting speed, rough surface is produced due to abrupt spark and that creates an uneven and multiple overlapped craters on the surface. At wire feed of 6 m/min, the better surface quality was achieved due to continuous optimum spark generation. Further increased wire feed to 8 m/min, the cutting speed rate increases and uneven spark produced from the wire melts unfairly, causing uneven craters. Further, due to this, the globules of debris on the surrounding of the crater cause the higher surface roughness.

Fig. 7.4 indicates the relationship between surface roughness and pulse on time for different values of pulse off time. The surface roughness decreases with increase in pulse off time irrespective of the pulse on time. This is due to the fact that increase in pulse off time causes to acquire uniform erosion of the material from the surface of the workpiece, otherwise nonuniform erosion of material takes place. Additionally, longer pulse off time furnishes good cooling effect and allows time to flush away the melted molten material and debris from the machined surface. Thus long pulse off time presents a better surface finish; the similar result has been reported by Lee and Li (2001).

The interaction effects of process parameters on surface roughness are exhibited in Fig. 7.5. In general, for any specified combination of pulse on time and wire feed, the surface roughness nonlinearly decreases with the increase in pulse on time. However, with the increase in wire feed from 4 to 6 m/min, the surface roughness decreases with the increase in pulse on time irrespective of pulse off time. On the other hand, with the increase in wire feed from 6 to 8 m/min, the surface roughness increases for lower values of pulse on time-pulse off time combination. The substantial surface quality deterioration is clearly evidenced for higher values of pulse on time-pulse off time combination. From Fig. 7.5, it is undoubtedly noticed that a combination of lower pulse on time (110 μs) and higher pulse off time (60 μs) with 6 m/min wire feed rate is found to beneficial for achieving better surface quality.

Fig. 7.6 depicts the SEM micrographs, which obviously indicates the presence of surface undulations, larger number of melted globule debris and craters. Fig. 7.6A represents the machined surface at lower pulse on time (110 μs) and pulse off time

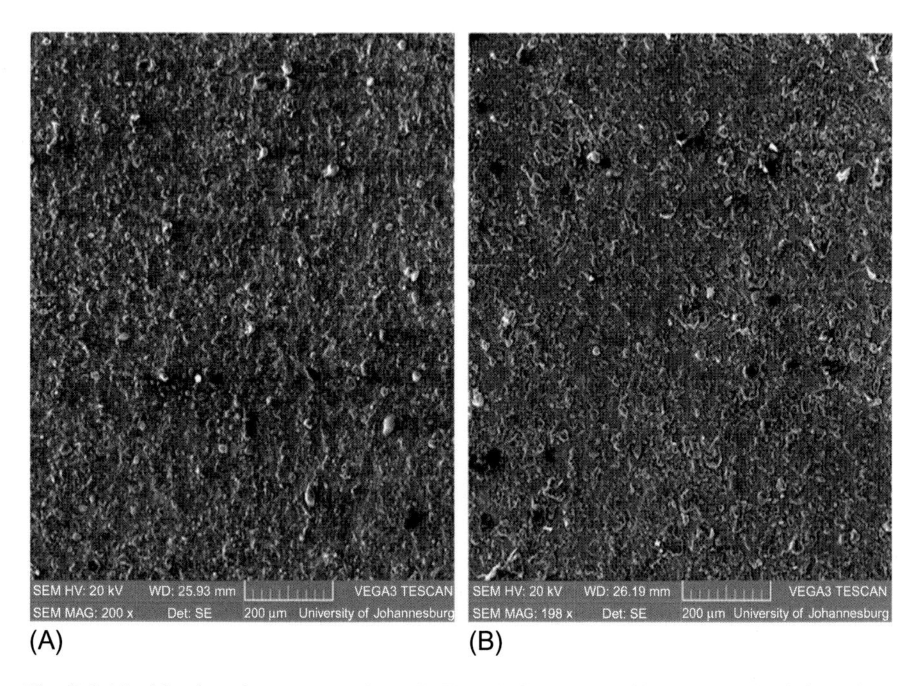

Fig. 7.6 Machined surface topography: (A) T_{on}—110 μs, T_{off}—40 μs, WF—6 m/min and (B) T_{on}—120 μs, T_{off}—60 μs, WF—8 m/min.

(40 μs), having a surface roughness of 1.26 μm. Fig. 7.6B illustrates the machined surface at higher pulse on time (120 μs) with higher pulse off time (60 μs) having a higher surface roughness of 2.80 μm. Higher pulse on time melts more amount of material, the molten material spells around the machined surface and thus forms larger globule of debris and causing higher surface roughness. At lower pulse on time, the material melting is minimal and the melted material is swept away due to average pulse off time (40 μs). Because of this, the smaller granular debris is formed on the machined surface and hence causes lower surface roughness.

7.4.2 *Analysis of material removal rate (MRR)*

Fig. 7.7 presents the effect of pulse on time and pulse off time on MRR. It is seen from the figure that an increase in pulse on time increases the MRR. The MRR incremental rate is more at lower pulse off time with higher pulse on time combination. This is because higher the pulse on time, larger will be the discharge energy and spark intensity leading to more amount of material removal. A combination of longer pulse on time with lower pulse off time also causes more sparking time and thus causing higher MRR. With the increase in pulse off time, the MRR gradually decreases, mainly due to reduced spark discharge time and less amount of material removal (Dhobe et al., 2013).

Fig. 7.8 shows the influence of pulse off time and wire feed on MRR. The MRR is decreased with increased pulse off time. Further, it is observed that the MRR is higher both at lower and higher levels of wire feed than the canter level of wire feed. This is because increase in wire feed increases the rate of cutting speed, which in turn increases the rate of spark discharge. More amounts of material melting and erosion takes place for higher spark discharge. Hence, MRR was more at higher wire feed and lower pulse off time.

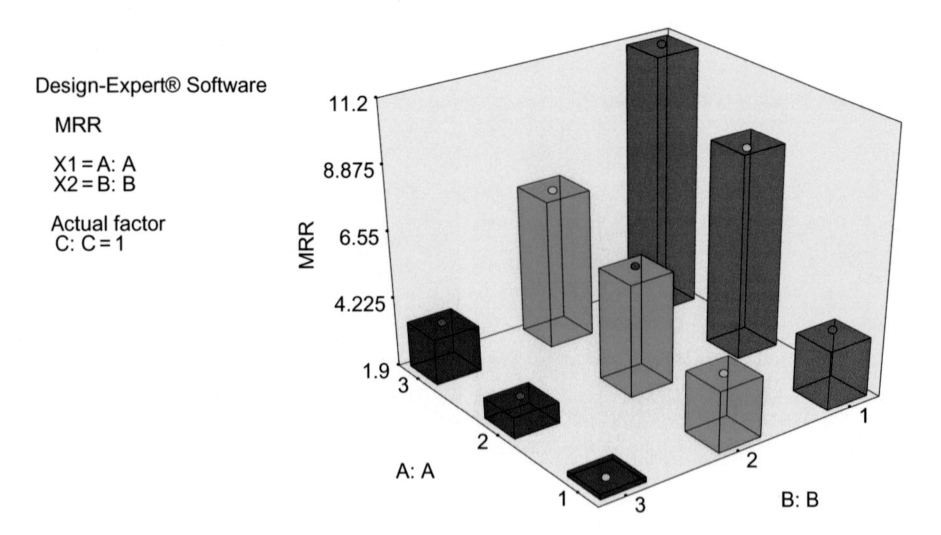

Fig.7.7 Effect of pulse on time and pulse off time against the MRR.

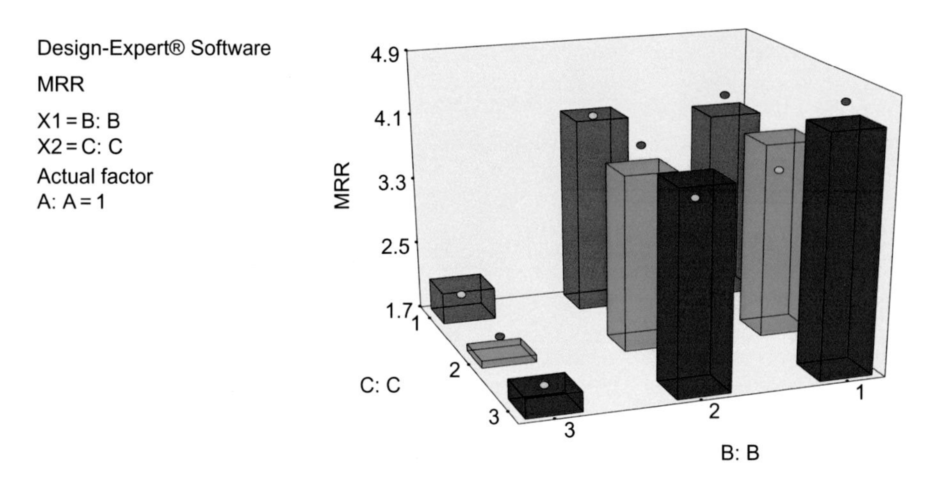

Fig. 7.8 The effect of pulse off time and wire feed against MRR.

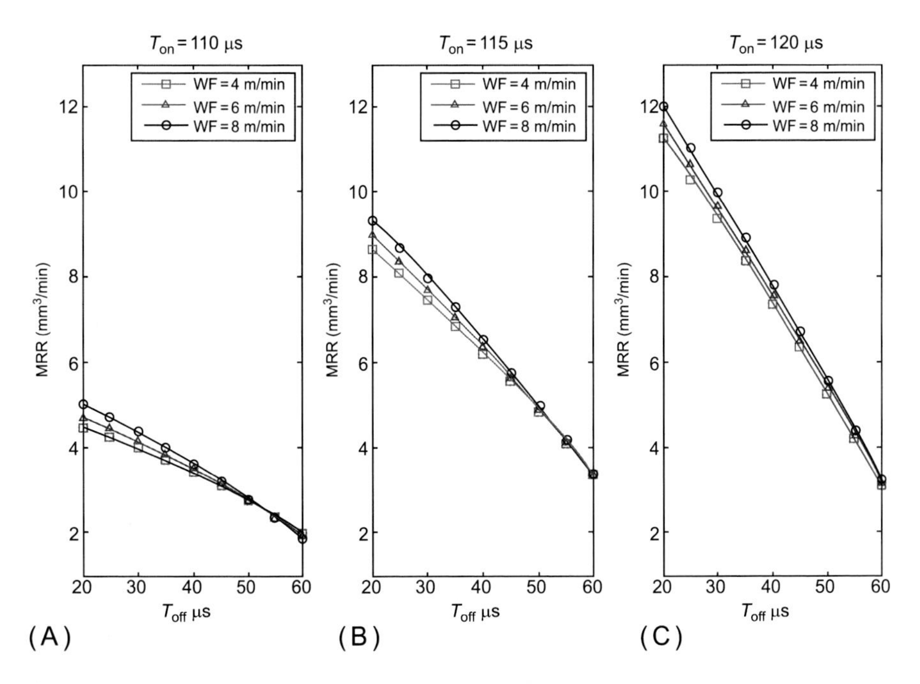

Fig. 7.9 Interaction effects of process parameters on MRR.

The variations of MRR are displayed in Fig. 7.9, which visibly demonstrates the linear behavior. Fig. 7.9 indicates that for any given value of pulse on time and wire feed, the MRR decreases with the increase in pulse off time; however, simultaneous increase in wire feed rate as well as pulse on time, the MRR increases. Hence, the combination of lower value of pulse off time along with higher values of pulse on time with wire feed rate is necessary for higher MRR.

7.4.3 Analysis of tool wear rate (TWR)

Fig. 7.10 illustrates the effect of pulse on time and pulse off time on TWR. TWR increases with the increase in pulse on time, which is due to high intense spark producing from the wire. The high electric discharge energy during longer pulse on time causes more heat dissipation and hence the wire rupture will be more due to excess thermal and increased discharge energy. As noticed in this figure, at lower pulse on and pulse off time, the TWR is found to be negligible. This is because the dissipated heat will vanish due to less sparking energy and less duration. Fig. 7.10 also reveals that the TWR decreased with increased pulse off time. The increased pulse off time allows a longer time to maintain the dissipated heat by sparking. The melted material flushed away from the machined surface also helps for the reduction in the heat.

Fig. 7.11 shows the influence of pulse off time and wire feed on TWR. It is observed that the TWR is higher at 6 m/min of wire feed rate for varying pulse off time. Also, TWR is greater at longer pulse off time (60 μs) and wire feed (8 m/min). Increased cutting speed forms larger crater on the wire surface caused by erosion of wire material. The dissipated heat is not dispersed sufficiently around the wire and work surface. The higher wire feed (8 m/min) with lower pulse off time (20 μs) combination has resulted in the lowest TWR. Because of higher cutting speed, continuous spark was generated at each interval of time and further the generated heat was nullified by the dielectric fluid for higher feed and lower pulse off time.

Fig. 7.12 presents the variations due to the interaction of pulse on time and pulse off time on tool wear rate for different wire feed rates. It is very interesting to note that for lower pulse on time of 110 μs, the tool wear rate linearly increases with the increase in pulse off time for any given wire feed rate; with further increase in wire feed from 4 to 6 m/min, the tool wear rate increases. However, with the increase in wire feed rate

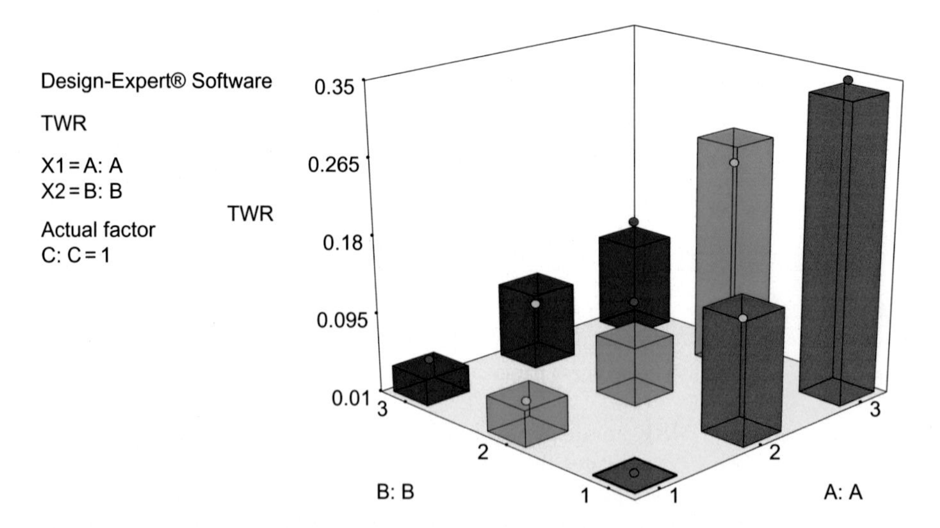

Fig. 7.10 The effect of pulse off time and wire feed against TWR.

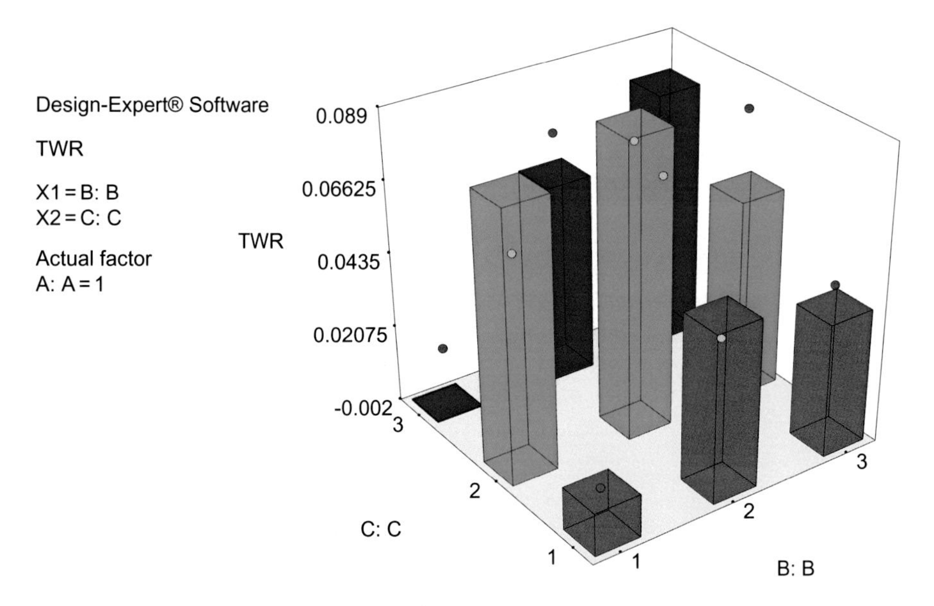

Fig. 7.11 The effect of pulse off time and wire feed against TWR.

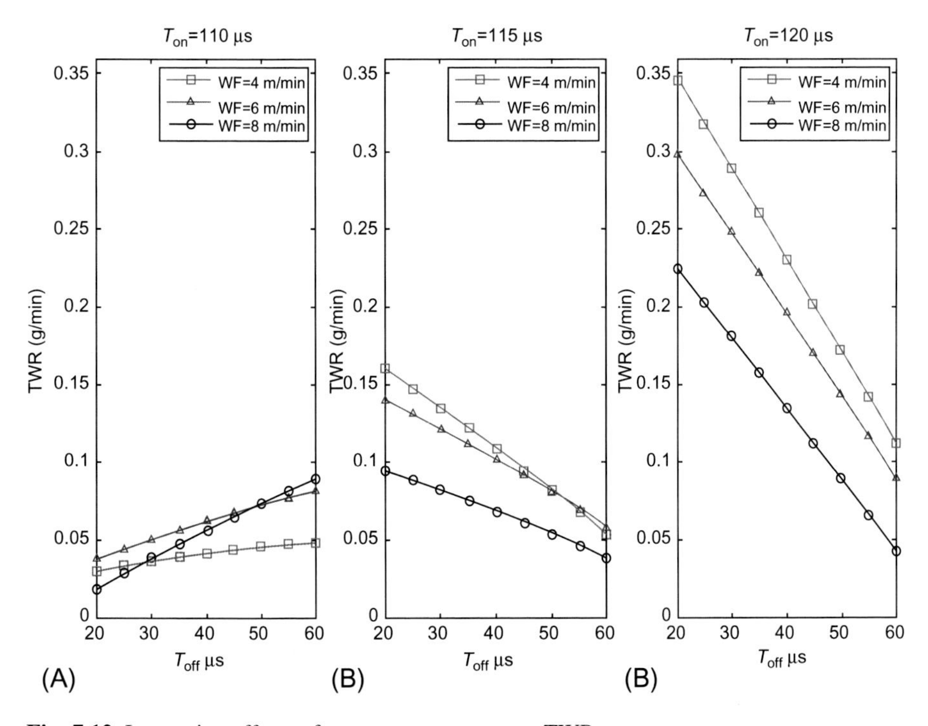

Fig. 7.12 Interaction effects of process parameters on TWR.

from 6 to 8 m/min, the tool wear rate decreases. On the contrary, for medium and higher values of pulse on time, tool wear rate linearly decreases with pulse off time, irrespective of the feed rate and exhibiting more or less similar behavior. Simultaneous increase in pulse on time along with wire feed rate provides the higher tool wear rate. Lower tool wear rate is observed for a combination of lower pulse on time, lower pulse off time with higher wire feed rate.

7.4.4 X-ray diffraction analysis

XRD analysis was performed on the machined surface of the sample in order to verify the presence of elements on the machined surface using XRD testing machine. XRD tests were carried out at a 2θ angle of 30 degrees to 80 degrees at a scan rate of 2 degrees/min. Fig. 7.13 shows the obtained XRD pattern for trial conditions. The formation of Fe_3O_4 (Ferrous oxide) and Fe_3C (Cementite) is clearly evidenced. Due to the disassociation of dielectric fluid and reaction with Fe, the Fe_3O_4 was formed on the machined surface. Cr and Cu elements are transferred from the wire surface to the machined surface during melting and evaporation. As can be seen from Fig. 7.13, the decrease in intensity, the formation of new peaks, and the increased Cr peak appeared at Trial 27 compared to Trial 5 and 15. This is because of higher pulse on time and wire feed rate. The XRD peak shows some widening of peaks, which is the indication of amorphous structure, possibly due to rapid quenching effect during machining. As reported by Bhattacharya et al. (2015), amorphous structure and formation of oxides and carbides increase the surface microhardness. In addition, the mechanical and physical properties such as toughness and wear resistance also improve due to increased hardness.

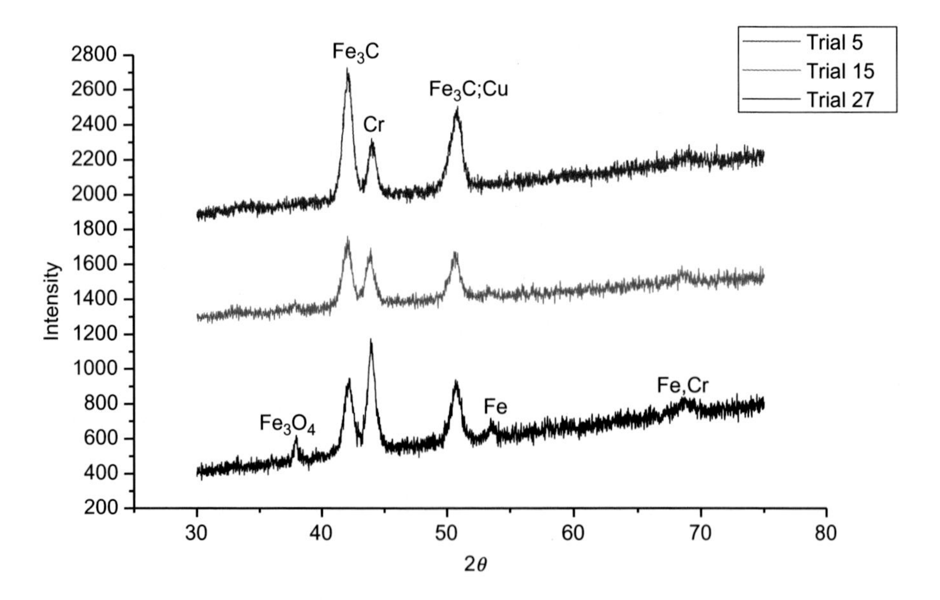

Fig. 7.13 X-ray diffraction analysis of WED-machined HCHCr steel.

7.4.5 Machined surface topography and EDX analysis

Fig. 7.14 shows the SEM images of HCHCr steel after WED machined. The micrograph of machined surface shows the deposited particle, globule of debris, craters, blowholes, and overlapped layered deposition. Fig. 7.14A shows the machined surface at pulse on time of 110 μs and Fig. 7.14B at pulse on time of 120 μs. More amount of resolidified layer, more amounts of globules, pores, crater, and blowholes are formed on the machined surface at higher pulse on time rather than lower pulse on time. This is because higher pulse on time leads to melting more amount of material and increased pressure in the molten causes entrapment of plasma gases and thus creates blowholes and craters on the machined surface. The machined surface consists of some foreign particles such as carbides, oxides, zinc, and increased carbon. The copper percentage was noticed on the machined surface by the energy dispersive spectroscopy (EDS) analysis (Fig. 7.15).

As seen from the plots (Figs. 7.5, 7.9, and 7.12.), there exists significant interaction between the process parameters on the proposed WEDM characteristics. Conversely, the behavior of the one WEDM characteristic is completely diverse from the other. For a given pulse of time, a combination of pulse on time and wire feed, which minimizes the surface roughness may not be suitable for minimizing the tool wear as well as maximizing the material removal rate. Hence, different combinations of pulse on time and wire feed are essential to simultaneously minimize the surface roughness as well as tool wear rate and to maximize the material removal rate at different pulse off time values. Hence, differential evolution (DE), a multiresponse optimization tool has been

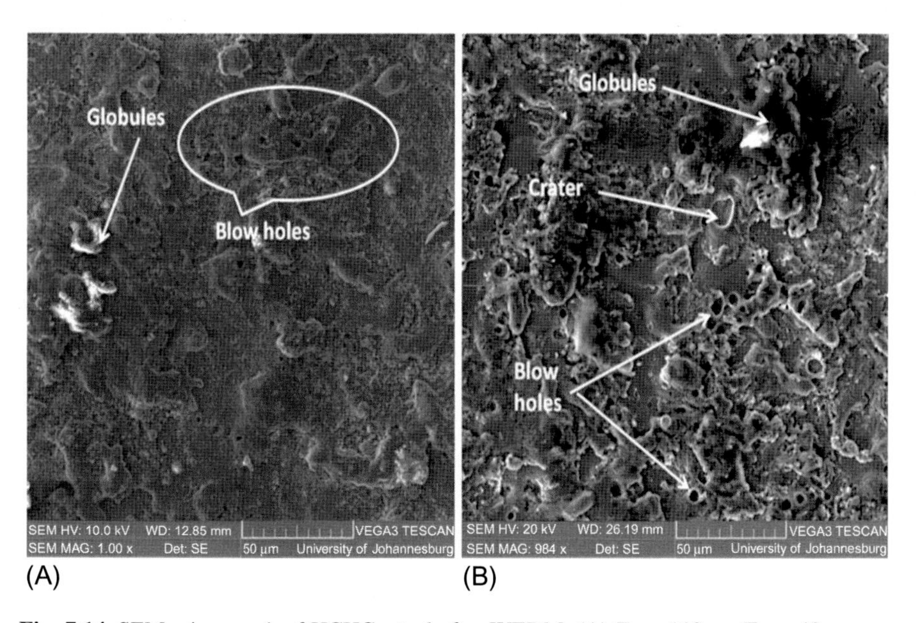

Fig. 7.14 SEM micrograph of HCHCr steel after WEDM: (A) T_{on}—110 μs, T_{off}—40 μs, WF—6 m/min, and (B) T_{on}—120 μs, T_{off}—60 μs, WF—8 m/min.

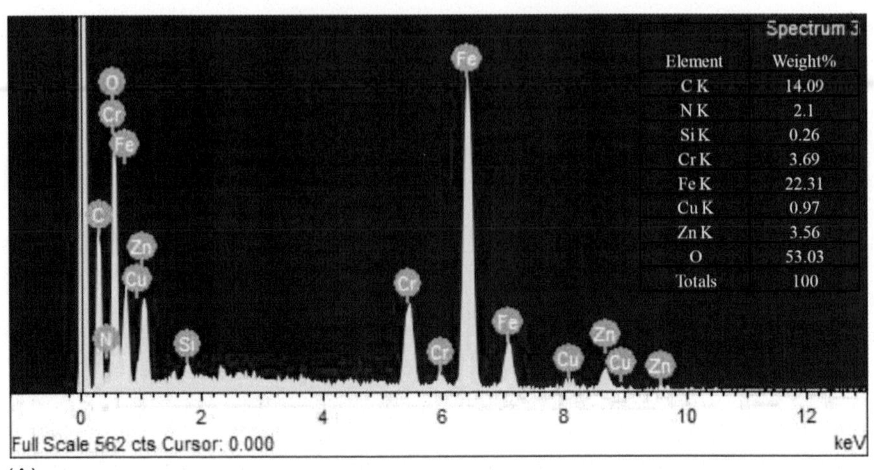

(A)

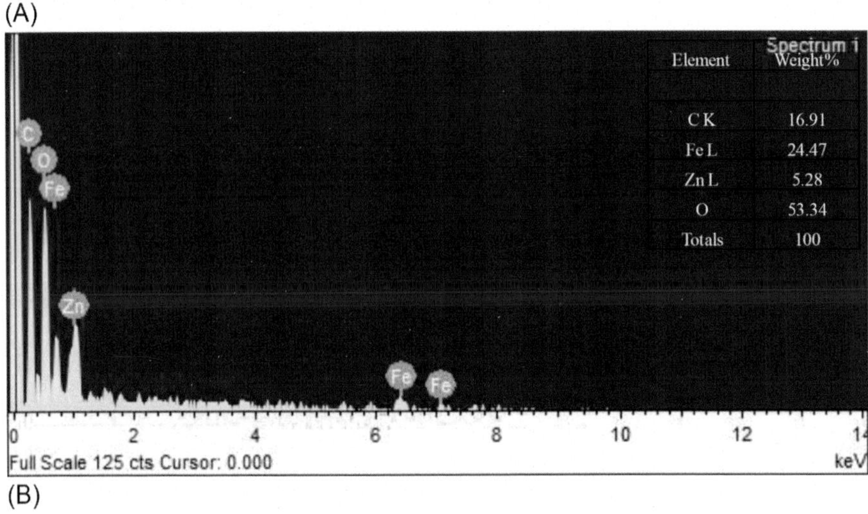

(B)

Fig. 7.15 EDS analyses of HCHCr steel after WEDM: (A) T_{on}—110 μs, T_{off}—40 μs, WF—6 m/min, and (B) T_{on}—120 μs, T_{off}—60 μs, WF—8 m/min.

employed for the determination of the best combination of pulse on time and wire feed for a specific pulse off time in the range 20–60 μs for simultaneous optimization of WEDM characteristics.

7.5 Differential evolution (DE) optimization

7.5.1 Overview

The differential evolution (DE) is an intelligent search algorithm appropriate for optimizing nonlinear and nondifferentiable functions, which has shown an outstanding performance on a huge multiplicity of standard and practical problems

(Price et al., 2005). A population of randomly generated and real-encoding candidate solutions evolves to an optimal solution through the reproduction operation and selection (Price et al., 2005).

The four important steps involved in DE are as follows:

Chromosomes: A chromosome is considered as an array that consists of a set of control parameters. The parameters are encoded using floating-point numbers and are set as elements in the chromosomes.

Population initialization: The primary DE population with population size (NP) candidate solutions is generated at random from the parameter domains as per the following:

$$x_{i,j}^{(0)} = p_j^{min} + r\left(p_j^{max} - p_j^{min}\right) \tag{7.10}$$

where $x_{i,j}^{(g)}$ is the jth process parameter of the ith individual at the gth generation; $i=1,.......,NP$; $j=1,.......,M$ and M is the number of process parameters defined in the required objective function; p_j^{max} and p_j^{min} are the upper and lower bounds of parameter j and r is a uniformly distributed random value over the range of [0, 1].

Reproduction operation: In DE, each parent x_i will produce one offspring u_i in every generation. The reproduction operator (DE/rand-to-best/1/bin) is given as:

$$u_i^g = x_i^g + K\left(x_{best} - x_i^g\right) + F\left(x_{r1}^g - x_{r2}^g\right) \tag{7.11}$$

where $r1, r2 = 1,......., NP$ are randomly selected number with $r1 \neq r2 \neq i$; x_{best} is the up-to-date best individual; K and F are scale coefficient of crossover and mutation, respectively; $K\left(x_{\text{best}} - x_i^g\right)$ and $F\left(x_{r1}^g - x_{r2}^g\right)$ play a role of crossover and mutation operation, respectively.

Selection strategy: A one-to-one replacement strategy is used for DE selection as per the following equation:

$$x_i^{g+1} = u_i^g \quad \text{if } f\left(u_i^g\right) < f\left(x_i^g\right);$$

Otherwise

$$x_i^{g+1} = x_i^g \tag{7.12}$$

In DE, the current best vector of the population can only be replaced by a better vector.

7.5.2 Development of DE optimization for WEDM characteristics

The constructed RSM-based models of surface roughness (Ra), material removal rate (MRR), and tool wear rate (TWR) are employed with the DE to find out the best combination values of pulse on time (T_{on}) and wire feed (WF) for a specified pulse off time (T_{off}) in the range 20–60 µs for simultaneously minimizing the surface roughness and tool wear rate and maximizing the tool wear rate.

In the current study, the objective function is defined as:

$$fit = \left[\frac{\mathrm{Ra}}{3.8} + \frac{\mathrm{TWR}}{0.05} + \left(1 - \frac{\mathrm{MRR}}{12}\right)\right] \tag{7.13}$$

Subject to the constraints: $110 < T_{on} < 120$ μs; $4 < \mathrm{WF} < 8$ m/min.

The DE simulation using RSM-based models is performed using MATLAB software (Math Works Incorporation, 2005) with a population size (NP) of 50 and maximum number of 300 generations. Scale coefficients for crossover (K) and mutation (F) were taken as 0.8 and 0.6, respectively. For each pulse-off time in the range 20–60 μs, 20 runs of DE optimization program were run and the best optimal responses and the corresponding pulse on time (T_{on}) and wire feed (WF) were recorded.

The optimal values of surface roughness (Ra), material removal rate (MRR), and tool wear rate (TWR) for various pulse off time (T_{off}) in the range 20–60 μs along with the corresponding optimal values of pulse on time (T_{on}) and wire feed (WF) are summarized in Table 7.5. As observed from Table 7.5, the optimal values of surface roughness (Ra) as well as material removal rate (MRR) decrease with the increased pulse off time (T_{off}) up to 50 μs and then both increase with pulse off time (T_{off}) beyond 50 μs. On the other hand, the optimal tool wear rate (TWR) increases with the increased pulse off time (T_{off}) in the range 20–50 μs and then decreases beyond 50 μs. From Table 7.5, it is also seen that the optimal pulse on time (T_{on}) remains constant, i.e., 110 μs (lower value), even if the pulse off time is varied from 20 to 50 μs, whereas the optimal pulse on time (T_{on}) increases with the increase in pulse off time (T_{off}) beyond 50 μs. However, the optimal wire feed is found to be

Table 7.5 Differential evolution (DE) optimization results for WEDM characteristics

		Optimal responses			Optimal control factors	
T_{off} (μs)	fit	Ra (μm)	MRR (mm^3/min)	TWR (g/min)	T_{on} (μs)	WF (m/min)
20	1.3921	1.612	5.011	0.0193	110	8
25	1.5948	1.55	4.699	0.0289	110	8
30	1.7973	1.499	4.362	0.0383	110	8
35	1.8348	1.3746	3.699	0.0391	110	4
40	1.8899	1.302	3.400	0.0415	110	4
45	1.9447	1.2418	3.077	0.0437	110	4
50	1.9993	1.1932	2.730	0.0456	110	4
55	2.1217	2.1428	4.220	0.0450	115	8
60	1.9764	2.2780	3.458	0.0330	116.5	8

8 m/min (higher) for the pulse off time in the range 20–30 μs, 4 m/min (lower) for the pulse off time in the range 30–50 μs, and again reaches the higher wire feed of 8 m/min for the pulse off time in the range 50–60 μs.

The DE optimization results illustrated in the current investigation are one such probable best solution.

7.6 Conclusions

The differential evolution (DE) optimization for simultaneous minimization of surface roughness (Ra) as well as tool wear rate (TWR) and the maximization of material removal rate (MRR) in wire electric discharge machining (WEDM) of HCHCr steel has been presented in this chapter. The minimum experiments were planned as per Taguchi's L_{27} orthogonal array (OA) and the quadratic models of WEDM characteristics, namely, surface roughness (Ra), material removal rate (MRR), and tool wear rate (TWR) were constructed based on response surface methodology (RSM) in terms of pulse on time (T_{on}), pulse off time (T_{off}), and wire feed (WF). The fitted second-order mathematical models were statistically tested through the analysis of variance (ANOVA) at 95% confidence interval. The fitness function was designed by mapping the surface roughness (Ra), material removal rate (MRR), and tool wear rate (TWR) for DE optimization. For every value of pulse off time (T_{off}) specified, the optimal values of pulse on time (T_{on}) and wire feed (WF) were determined through DE simulation. The following are concluded from the current experimental studies and subsequent analysis:

- There exists remarkable interaction between the identified input parameters on the planned WEDM characteristics and accordingly justifying the exercise of RSM.
- The surface roughness is higher at lower wire feed and further increase in wire feed decreases the surface roughness; again it increases at higher wire feed rate. A combination of lower pulse on time (110 μs), higher pulse off time (60 μs) with 6 m/min wire feed rate is found to beneficial for achieving better surface quality.
- For any specified value of pulse on time and wire feed, the MRR decreases with increased pulse off time; however, the concurrent increase in wire feed rate and pulse on time, the MRR increases.
- Simultaneous increase in pulse on time with wire feed gives the higher tool wear rate. Lower tool wear rate is observed for a combination of lower pulse on time, lower pulse off time with higher wire feed.
- DE optimization results reveal that the optimal values of surface roughness (Ra) and material removal rate (MRR) decrease with the increased pulse off time (T_{off}) up to 50 μs and then increase with pulse off time (T_{off}). Alternatively, the optimal tool wear rate (TWR) increases with increased pulse off time (T_{off}) in the range 20–50 μs and afterward decreases.
- The optimal pulse on time (T_{on}) remains constant at lower value of 110 μs for a pulse off time in the range 20–50 μs, whereas it increases with the increase in pulse off time (T_{off}) beyond 50 μs. The optimal wire feed is at higher value of 8 m/min for the pulse off time in the range 20–30 μs, lower value of 4 m/min for the pulse off time in the range 30–50 μs, and yet again reaches the higher wire feed of 8 m/min for the pulse off time in the range 50–60 μs.

References

Bhattacharya, A., Batish, A., Bhatt, G., 2015. Material transfer mechanism during magnetic field-assisted electric discharge machining of AISI D2, D3 and H13 die steel. Proc. Inst. Mech. Eng. B J. Eng. Manuf. 229 (1), 62–74.

Coldwell, H., Woods, R., Paul, M., Koshy, P., Dewes, R., Aspinwall, D., 2003. Rapid machining of hardened AISI H13 and D2 moulds, dies and press tools. J. Mater. Process. Technol. 135 (2–3), 301–311.

Design-Expert, 2016. Design-Expert®, Version 7.0. Stat-Erase, Inc, Minneapolis, MN.

Dhobe, M.M., Chopde, I.K., Gogte, C.L., 2013. Investigations on surface characteristics of heat treated tool steel after wire electro-discharge machining. Mater. Manuf. Process. 28 (10), 1143–1146.

Esmat, R., Hossein, N., Saeid, S., 2009. GSA: a gravitational search algorithm. Inf. Sci. 179, 2232–2248.

Gaitonde, V.N., Karnik, S.R., 2012. Minimizing burr size in drilling using artificial neural network (ANN)-particle swarm optimization (PSO) approach. J. Intell. Manuf. Syst. 23 (5), 1783–1793.

Gaitonde, V.N., Karnik, S.R., Achyutha, B.T., Siddeswarappa, B., 2008a. Genetic algorithm based burr size minimization in drilling of AISI 316 L stainless steel. J. Mater. Process. Technol. 197 (1–3), 225–236.

Gaitonde, V.N., Karnik, S.R., Siddeswarappa, B., Achyutha, B.T., 2008b. Integrating Box-Behnken design with genetic algorithm to determine the optimal parametric combination for minimizing burr size in drilling of AISI 316 L stainless steel. Int. J. Adv. Manuf. Technol. 37 (3–4), 230–240.

Gaitonde, V.N., Karnik, S.R., Davim, J.P., 2012a. Optimal MQL and cutting conditions determination for desired surface roughness in turning of brass using genetic algorithms. Mach. Sci. Technol. 16 (2), 304–320.

Gaitonde, V.N., Karnik, S.R., Davim, J.P., 2012b. Minimising burr size in drilling: integrating response surface methodology with particle swarm optimisation. In: Davim, J.P. (Ed.), Mechatronics and Manufacturing Engineering: Research and Development. Woodhead Publishing Limited, Cambridge, UK, pp. 259–292 (Chapter 7).

Gaitonde, V.N., Karnik, S.R., Davim, J.P., 2012c. Application of particle swarm optimization for achieving desired surface roughness in tungsten-copper alloy machining. In: Davim, J.P. (Ed.), Computational Methods for Optimizing Manufacturing Technology: Models and Techniques. IGI Global Science, pp. 144–161 (Chapter 6).

Hascalik, A., Caydas, U., 2007. Electrical discharge machining of titanium alloy (Ti–6Al–4 V). Appl. Surf. Sci. 253 (22), 9007–9016.

Jomaa, W., Fredj, N. Ben, Zaghbani, I., Songmene, V., 2011. Non-conventional turning of hardened AISI D2 tool steel. Int. J. Adv. Mach. Form. Oper. 3 (2), 1–41.

Karnik, S.R., Gaitonde, V.N., Davim, J.P., 2007. Integrating Taguchi principle with GA to minimize burr size in drilling of AISI 316 L stainless steel using ANN model. Proc. Inst. Mech. Eng., IMech E, J. Eng. Manuf. 221, 1695–1704.

Karnik, S.R., Gaitonde, V.N., Basavarajappa, S., Davim, J.P., 2013. Multi-response optimization in drilling of glass epoxy polymer composites using simulated annealing approach. Mater. Sci. Forum 766, 123–141.

Lee, S.H., Li, X.P., 2001. Study of the effect of machining parameters on the machining characteristics in electrical discharge machining of tungsten carbide. J. Mater. Process. Technol. 115 (3), 344–358.

Lodhi, B.K., Agarwal, S., 2014. Optimization of machining parameters in WEDM of AISI D3 steel using Taguchi technique. Procedia CIRP 14, 194–199.

Manjaiah, M., Narendranath, S., Basavarajappa, S., Gaitonde, V.N., 2014. Some investigations on wire electric discharge machining characteristics of titanium nickel shape memory alloy. Trans. Nonferrous Metals Soc. China 24 (10), 3201–3209.

Manjaiah, M., Narendranath, S., Basavarajappa, S., Gaitonde, V.N., 2015. Effect of electrode material in wire electro discharge machining characteristics of $Ti_{50}Ni_{50-x}Cu_x$ shape memory alloy. Precis. Eng. 41, 68–77.

Manjaiah, M., Narendranath, S., Basavarajappa, S., Gaitonde, V.N., 2016a. Influence of process parameters on material removal rate and surface roughness in WED-machining of $Ti_{50}Ni_{40}Cu_{10}$ shape memory alloy. Int. J. Mach. Mach. Mater. 18 (1/2), 36–53.

Manjaiah, M., Laubscher, R.F., Narendranath, S., Basavarajappa, S., Gaitonde, V.N., 2016b. Evaluation of wire electro discharge machining characteristics of $Ti_{50}Ni_{50-x}Cu_x$ shape memory alloys. J. Mater. Res. 31 (2), 1801–1808.

Mata, F., Garrido, I., Tejero, J., Gaitonde, V.N., Karnik, S.R., Davim, J.P., 2013. Surface roughness minimization in turning PEEK-CF30 composites with TiN cutting tools using particle swarm optimization. Mater. Sci. Forum 766, 109–122.

Math Works Incorporation, 2005. MATLAB User Manual, Version 7.1, R 14, Natick, MA.

Mir, M.J., Sheikh, K., Singh, B., Malhotra, N., 2012. Modeling and analysis of machining parameters for surface roughness in powder mixed EDM using RSM approach. Int. J. Eng. Sci. Technol. 4 (3), 45–52.

Montgomery, D.C., 2003. Design and Analysis of Experiments. John Wiley and Sons, New York.

Myers, R.H., Montgomery, D.C., Anderson-cook, C.M., 2009. Response Surface Methodology: Process and Product Optimization Using Designed Experiments. John Wiley and Sons, Inc., New Jersey.

Narendranath, S., Manjaiah, M., Basavarajappa, S., Gaitonde, V.N., 2013. Experimental investigations on performance characteristics in wire EDM of $Ti_{50}Ni_{42.4}Cu_{7.6}$ shape memory alloy. Proc. Inst. Mech. Eng. B J. Eng. 227 (8), 1180–1187.

Novotny, P.M., 2001. Tool and Die Steels. Encyclopedia Mater. Sci. Technol. 9384–9389.

Price, K.V., Storn, R.M., Lampinen, J.A., 2005. Differential Evolution – A practical Approach to Global Optimization. Springer.

Purohit, R., Rana, R.S., Dwivedi, R.K., Banoriya, D., Singh, S.K., 2015. Optimization of electric discharge machining of M2 tool steel using grey relational analysis. Mater. Today: Proc. 2 (4–5), 3378–3387.

Ramakrishnan, R., Karunamoorthy, L., 2006. Multi response optimization of wire EDM operations using robust design of experiments. Int. J. Adv. Manuf. Technol. 29 (1-2), 105–112.

Singh, V., Pradhan, S.K., 2014. Optimization of WEDM parameters using Taguchi technique and response surface methodology in machining of AISI D2 steel. Procedia Eng. 97, 1597–1608.

Index

Note: Page numbers followed by *f* indicate figures, and *t* indicate tables.

A

Abaqus software, 61, 71
Adaptive network-based fuzzy inference system (ANFIS), 189–190
ALE analysis. *See* Arbitrary Lagrangian-Eulerian (ALE) analysis
Aluminum alloys, 125, 132
Amdahl's law, 2–3, 16–17, 20–21, 26
ANSYS Fluent, 153–154
Arbitrary Lagrangian-Eulerian (ALE) analysis, 98, 139, 179
Artificial neural network (ANN)
  hard machining, 186–189
  hot rolling process, 90

B

Blank holder force (BHF), 52
Boundary conditions, FSW, 137
  CEL analysis, 161, 161*f*
  Eulerian analysis, 153–154, 153*f*
  Lagrangian analysis, 147
B-splines, 63

C

Central processing unit (CPU) time, 1
Conjugate gradient (CG) iterative solvers, 1–2
Constitutive laws, 130
Contact interaction, FSW
  CEL analysis, 160–161
  Lagrangian analysis, 146–147, 146*f*, 147*t*
Control surface method, 53
Convective heat transfer, 153–154, 160–161
Coordinate systems (CSs), 33
Cost function formulation, sheet metal forming process, 60
Coulomb's law of friction, 134, 160–161
Coupled Eulerian-Lagrangian (CEL) analysis, FSW, 139
  boundary conditions, 161, 161*f*
  contact interaction, 160–161
  geometric modeling and material model, 158–159, 159*t*, 159*f*
  governing equations, 161–163, 163*f*
  mesh generation, 159–160, 159–160*f*, 161*t*
  results and discussion, 163–164, 163–164*f*
  split operator for, 163*f*

D

DD3IMP software, 61, 64
Deformation
  elastic-plastic, 129
  plastic, 130
Design of Experiments (DoE) techniques, 54
Differential evolution (DE) optimization, 216–219
Direct curvature method (DCA), 53
Direct equation solver, parallelization of. *See* Parallel direct solver
Discretization process, 137–138
Displacement adjustment (DA) method, 53

E

Elastic-plastic deformation, 129
Element, 96
Element stiffness equation, 96
Energy dispersive spectroscopy (EDS) analysis, 215–216
Eulerian analysis, FSW, 139
  boundary condition, 153–154, 153*f*
  geometric modeling, 153–154, 153*f*
  governing equation, 156
  heat generation, 156–157
  material model, 154–155, 155*t*
  mesh generation and solver, 157
  results and discussion, 157–158, 157*f*
Eulerian approach, 98

F

Faro gauge, 203*f*
Feed-forward control, 37*f*, 39–43, 47
Feed-forward predictive control, 36–39

Finite element analysis (FEA)
hot rolling
advantages, 85–86
arbitrary Lagrangian-Eulerian approach, 98
deformation analysis, 107–109
discretization of domain, 96
element stiffness equation, 96
Eulerian approach, 98
FEA software, 88–89
global stiffness matrix, 96
Lagrangian approach, 97–98
online application, 90
research works, 86–87
solving the equations, 97
temperature inhomogeneity measurement, 89–90
thermal analysis, 109–110
three-dimensional model, 87–88
two-dimensional model, 87–88
software, 60–61
Finite element method (FEM), 84–85, 137–138, 149, 178–179
Finite Element Model Updating (FEMU)
vs. metamodels, 74–76
strategies, 59–60
Finite element simulations, parallel direct solver
evaluation parameters, 16
structure, 6–13
test cases, 14–16
Flat hot rolling, 83
Flat rolling, 83
Flow stress model, 183–184
Fluid structure interaction (FSI), 139
Force descriptor method (FDM), 52–53
Friction coefficient, during rolling, 93–94
Friction stir welding (FSW), 131*t*
ALE analysis, 139
boundary conditions, 137
capabilities of software, 139, 140*t*
CEL analysis, 139
boundary conditions, 161, 161*f*
contact interaction, 160–161
geometric modeling and material model, 158–159, 159*t*, 159*f*
governing equations, 161–163, 163*f*
mesh generation, 159–160, 159–160*f*, 161*t*
results and discussion, 163–164, 163–164*f*
complexity level and analysis type, 128–129
Eulerian analysis, 139
geometric modeling and boundary condition, 153–154, 153*f*
governing equation, 156
heat generation, 156–157
material model, 154–155, 155*t*
mesh generation and solver, 157
results and discussion, 157–158, 157*f*
FEM, 137–138, 149
geometric modeling and assembly, 129–130
heat transfer in, 136*f*
invention of, 125
Johnson-Cook model, 130–132, 132*t*
Lagrangian analysis, 138, 141
boundary conditions and assumptions, 147
contact interaction, 146–147, 146*f*, 147*t*
force and torque evolution, 152, 152*f*
geometric modeling and material model, 141, 142*f*, 142*t*
governing equation, 148–149
mesh generation, 141–146, 143–145*f*, 144*t*
solvers and iterative method, 149, 150*f*
temperature and plastic strain distribution, 149–152, 151*f*
lexicon, 126
material properties and constitutive equation, 130–132
mechanical interactions, 133–135, 133*f*, 134–136*t*
mesh generation, 137–138
modeling, 127–128
nomenclature, 127*f*
nonconsumable tool, 125–126
requirement and complexities, 127–128
Sheppard and Wright model, 132, 133*t*, 141, 142*t*, 154
stages, 126*f*
thermal interaction, 135–137, 136*f*
two-dimensional model, 128–129
Fuzzy logic models, hard machining, 189–191

G

Gaussian process regression, 59
Generalized reduced gradient (GRG) method, 62
Geometric models, FSW, 129–130
  Eulerian analysis, 153–154, 153*f*
  Lagrangian analysis, 141, 142*f*, 142*t*
Global stiffness matrix, 97
Governing equations, FSW
  CEL analysis, 161–163, 163*f*
  Eulerian analysis, 156
  Lagrangian analysis, 148–149

H

Hard machining
  computing and statistical methods
    artificial neural networks, 186–189
    fuzzy logic models, 189–191
    response surface methodology, 191–193
  hard drilling, 176
  hard milling, 175–176
  hard turning, 171–174
  plastic deformation, 176
  surface layers, 176–177
  surface quality, 176
  thermal models, 177–178
  thermomechanical models
    arbitrary Lagrangian-Eulerian model, 179
    finite elements method, 178–179
    with flow stress, 181–184
    microstructure-related phenomena, 184–185
    smoothed particle hydrodynamics model, 180
    two-dimensional numerical model, 181
Heat generation, FSW, 156–157
Heat transfer
  coefficients, parameters for, 147*t*
  convective, 153–154, 160–161
  in FSW, 136*f*
Hot rolling
  boundary conditions
    mechanical boundary, 103, 104*f*
    thermal boundary, 103
  computing environment, 110–112
  constitutive equation
    governing, 104–105
    for material model, 94–95, 107
  equilibrium equations, 104–105
  finite element analysis
    advantages, 85–86
    ALE approach, 98
    deformation analysis, 107–109
    discretization of domain, 96
    element stiffness equation, 96
    Eulerian approach, 98
    FEA software, 88–89
    global stiffness matrix, 96
    Lagrangian approach, 97–98
    online application, 90
    research works, 86–87
    solving the equations, 97
    temperature inhomogeneity measurement, 89–90
    thermal analysis, 109–110
    three-dimensional model, 87–88
    two-dimensional model, 87–88
  flat hot rolling, 83
  mesh sensitivity study, 112
  mesh update, 102
  rolling load and rolling torque, 118–119
  solution methods, 98–99
  solver, type of, 110
  steady-state analysis
    initial geometry, 100–101
    mesh generation, 101–102
  stress and strain distribution, 116–118
  temperature distributions, 112–116
  thermomechanical analysis, 86–91
  unsteady-state analysis
    initial geometry, 100–101
    mesh generation, 102
  work-roll and workpiece
    deformation, 91
    friction coefficient, 93–94
    interfacial heat transfer coefficient, 92
    interobject relation, 91
  yield function equation, 105
Hybrid model, 135

I

Interfacial heat transfer coefficient, 92
Interpolation function, 96
Iterative method, FSW, 149, 150*f*

Iterative solver. *See also* Conjugate gradient (CG) iterative solvers
  accuracy, 18–19, 26

J

Johnson-Cook model, 130–132, 132*t*

K

Kriging process regression, 59

L

Lagging zone, 83
Lagrangian analysis, FSW, 138, 141
  boundary conditions and assumptions, 147
  contact interaction, 146–147, 146*f*, 147*t*
  force and torque evolution, 152, 152*f*
  geometric modeling and material model, 141, 142*f*, 142*t*
  governing equation, 148–149
  mesh generation, 141–146, 143–145*f*, 144*t*
  solvers and iterative method, 149, 150*f*
  temperature and plastic strain distribution, 149–152, 151*f*
Lagrangian approach, 97–98
Lagrangian elastic model, 88–89
Leading zone, 83
Levenberg-Marquardt (L-M) method, 62

M

Machined surface topography, 215–216
Material removal rate (MRR), 210–211
Mechanical interactions, FSW, 133–135, 133*f*, 134–136*t*
Mesh generation, FSW, 137–138
  CEL analysis, 159–160, 159–160*f*, 161*t*
  Eulerian analysis, 157
  Lagrangian analysis, 141–146, 143–145*f*, 144*t*
Mesh refinement study, 143–145, 143–145*f*
Mesh sensitivity study, 112
Microstructure-related phenomena, hard machining, 184–185
Multiple linear regression (MLR), 57–58
Multistage assembly processes (MAPs)
  assembly operations and locating scheme, 44*t*
  case study, 43–45
  deviation propagation, 35*f*
  feed-forward control, 37*f*, 39–43, 47
  feed-forward predictive control, 36–39
  flow chart, 35*f*
  product and process design information, 44*t*
  state-space model, 33–36
  two-dimensional model, 33*f*

N

Nelder-Mead (N-M) Simplex algorithm, 62
Neutral plane, 83
Nodes, 96
Non-Uniform Rational B-Splines (NURBS), 63

O

OpenMP, parallel direct solver by, 13, 26–29
Optimal actuator placement, 42–43, 47–49
Optimal inspection/actuator placement
  actuator placement, 39–42
  constrained optimal control action, 48–49
  cost index, 42
  feed-forward control, 40*f*
  final state after control implementation, 49
  inspection stations placement, 42–43
  unconstrained optimal control action, 47–48
Optimal inspection placement, 39–42
Orthogonal array, 200, 202*t*

P

Parallel direct solver
  accuracy, 18–19
  advantages, 2
  benchmark test case, 14
  core loop, 8–13
  diagonal positions reduction, 11–12
  evaluation parameters, 16
  factorization and reduction, 7
  main loop, 8
  off-diagonal positions reduction, 11
  performance/efficiency, 19–25
  resistance welding test case, 15
  right-hand side vector reduction, 11–13
  skyline matrix storage, 2
  solution time, 17–18
  source code, 13, 26–29
  utilization as preconditioner, 13
Partial differential equation (PDE), 84–85

Penalty contact algorithm, 133, 133*f*, 160–161
Perzyna's viscoplasticity model, 154
Plastic deformation, 130
Polynomial regression model, sheet metal forming process, 58
Pressure-velocity coupling, 157
PRESTO algorithm, 157

R

Recurrent neural network (RNN) model, 188
Remeshing technique, 137–138, 141, 146
Residual stresses, 181
Resistance welding test case, parallel direct solver, 15
Response surface methodology (RSM), 55–59, 204–205
  hard machining, 191–193
Roll bite angle, 83
Roll chilling effect, 88–89, 114–115
Roll gap, 83
Rolling
  definition, 83
  deformation zone, 83
  schematic diagram, 84*f*
  three-dimensional representation, 84*f*
Roll-workpiece interface
  deformation, 91
  friction coefficient, 93–94
  interfacial heat transfer coefficient, 92
  interobject relation, 91

S

Shape function, 96
Shear friction model, 135
Sheet metal assembly process, 33*f*
Sheet metal forming process
  cost function formulation, 60
  experimental analysis, 61
  FEA software, 60–61
  FEMU *vs.* metamodels, 74–76
  finite element model updating, 59–60
  geometry parameterizations, 51–52
  metamodeling optimization
    modeling strategies and evaluation, 69–76
    multiple linear regression, 57–58
    polynomial regression, 58
    universal kriging, 59
  optimization algorithms, 62, 76–77
  parameterization strategies, 62–63, 78
  sensitivity analysis, 69
  springback compensation strategies, 52–55
  U-rail study, 63–68
Sheppard-Wright model, 132, 133*t*, 141, 142*t*, 154
Skyline matrix storage, parallel direct solver, 2
Smooth displacement adjustment (SDA) method, 53
Smoothed particle hydrodynamics (SPH) model, 180
Soft computing methods, for hard machining
  artificial neural networks, 186–189
  fuzzy logic models, 189–191
  response surface methodology, 191–193
Solvers method, FSW, 149, 150*f*
Springback compensation strategies
  direct curvature method, 53
  displacement adjustment method, 53
  flow chart, 56*f*
  force descriptor method, 52–53
  optimal compensation factors and adjustments, 53
  optimization algorithms, 54–55, 62, 76–77
  parameterization strategies, 62–63, 78
  Robust optimization and metamodels, 54
Standard gauss elimination
  backward substitution, 6
  factorization of system matrix, 4–5, 5*f*
  right-hand side vector division, 6
  right-hand side vector reduction, 4–5
  skyline matrix storage, 3
State-space model approach
  stage-level model, 33–34
  system-level model, 34–36
Statistical process control (SPC), 32
Steady-state analysis, hot rolling
  initial geometry, 100–101
  mesh generation, 101–102
Surface roughness (Ra), 208–210

T

Thermal interaction, FSW, 135–137, 136*f*
Tool wear rate (TWR), 212–214
Transformation-induced plasticity (TRIP), 184
T-splines, 63

U

Universal kriging, 59
Unsteady-state analysis, hot rolling
  initial geometry, 100–101
  mesh generation, 102
U-rail test
  conditions, 65*t*
  design variables range, 67*t*, 68
  dimensions and design variables, 63, 64*f*
  orthogonality and planarity of surfaces, 66–68
  parameterization, 65–66, 66*f*

W

Weak form, 96
Wire electric discharge machining (WEDM)
  differential evolution, 217–219
  EDX analysis, 215–216
  experimental work, 201–204
  machined surface topography, 215–216
  material removal rate analysis, 210–211
  response surface methodology, 204–205
  surface roughness analysis, 208–210
  tool wear rate analysis, 212–214
  X-ray diffraction analysis, 214
Wire feed (WF)
  against material removal rate, 210, 211*f*
  against surface roughness, 207*f*, 208
  against tool wear rate, 212, 212*f*

Z

Zener-Hollomon equation, 132
Zener-Hollomon parameter, 94–95